Design
Management

Peter Gorb   *Editor*

Design
Management

Papers from the
London Business
School

VNR   VAN NOSTRAND REINHOLD   *New York*

An ADT Press book

*Copyright © London Business School
Design Management Unit 1990*

ISBN 0-442-30363-7

*Published in the U.S.A. by
Van Nostrand Reinhold
115 Fifth Aveenue
New York, New York 10003*

*This book was designed and produced by
Architecture Design and
Technology Press
128 Long Acre
London WC2E 9AN
A Division of Longman Group UK Ltd*

Designer: Jonathan Moberly

First published in this form 1990

*Typeset by Discript, London
Printed in Great Britain by
The Bath Press*

Contents

Peter Gorb

Foreword

The literature on the relatively new subject of design management is comparatively rare. This collection of papers is only the second of its kind to be published. Like the first collection it had its origins in a seminar series which is run regularly at the London Business School. The papers in this volume were presented between October 1987 and June 1989.

The teaching of design at a postgraduate business school was pioneered at the London Business School. It began as far back as 1976 and a commitment was made to the Design Management Unit at the school in 1982. The unit now has full-time staff, undertaking teaching research, and doing what they can to generate wider interest in the subject.

The evening seminars at the school attempt to explore design management issues of general interest to both the management and the design communities. By gathering together in this book the papers from those seminars it is hoped to reach a wider audience, not only of students of management and design, but also working practitioners in both fields.

The papers fall naturally into two parts, and the book is divided accordingly. In Part I, entitled 'Viewpoints', the authors are concerned to explore the subject from their own professional and personal standpoints; but ultimately with very little conflict of opinion. So James Pilditch, with a deep experience of the commercial links established by design consultants, reaches many similar conclusions to Ivor Owen who approaches the subject from the viewpoint of an industrialist now in charge of national action as director of the Design Council. Similarly Lord Gowrie deals with some broad issues relating to the social psychology of design, a subject which Stephen Bayley also illuminates a paper which anticipates the exhibition which opened the new Design Museum. Finally, Angela Dumas and Allan Whitfield describe some of the strategic questions to which their research into the organization of design in industry has led them. My own paper which concludes this section is a polemic which argues some of the same considerations.

Part II of the book deals with case histories in which leading practitioners describe how design is managed in their organizations. Jane Priestman and Raymond Turner write as directors of design in their respective organizations; Sir Terence Beckett, John Clothier, Sir Brian Corby, Sir Ralph Halpern, Derek Lovelock and Geoffrey Maddrell all write as chief executives. The companies vary widely, although it is interesting to note that all of them deal either directly or partially with the general public. It is I suppose a truism that the closer an organization is to the consumer the greater is its commitment to design.

Certainly that commitment is very evident as a common theme throughout all of these papers and I should like to thank all of the authors for their contributions, given first as a seminar paper and secondly as written text for this book. They received no payment for the first and no royalties for the second. All royalties are given into the funds of the London Business School to further the cause of design management.

As to design management, it is perhaps prudent to ensure that before dipping into the rich collection of views and explanations to be found among these papers, the reader has some view of what design management is, what it does and what it is worth. The introduction which follows attempts an answer to those questions.

*Peter Gorb is Senior Fellow in
Design Management at the
London Business School
where he pioneered the
teaching of design to
managers. He moved to the
School after a career as a
line manager in industry. He
has published widely in the
field of design.*

Introduction: What is Design Management?

Design management as an idea has been around for over a quarter of a century. The Royal Society of Arts made its first Design Management awards in 1966. Since then the two words, design and management, have been used to convey a multiplicity of meanings. In 1979 I was able to describe four distinct ways in which the words were used, each meaning shading into the next.[1]

Multiplicity of meaning is not necessarily bad. It implies richness of thought and behaviour in a new and growing field. However, in the past decade as corporations have increasingly taken action to use design to improve performance, the time to codify design management has perhaps arrived.

What this paper attempts to do is to drive out all but one definition; to say what design management is, who does it, and what it is worth. But before doing that it is important to say what it is not.

Design management is not the process of managing a design consultancy or practice, either within or outside a corporation. It is not the education of designers about the importance of the management world;

or the reverse: educating managers about design. All these are important activities; indeed they are relevant, preliminary, and necessary to the effective practice of design management. But they are something else. Nor is design management synonymous with product development, or facilities management, or identity management. All of these are important aspects of the wider activity, but only aspects of it.

DEFINITIONS

Having cleared the ground let us attempt a definition. Design management is the effective deployment by line managers of the design resources available to an organization in the pursuance of its corporate objectives. It is therefore directly concerned with the organizational place of design, with the identification of specific design disciplines which are relevant to the resolution of key management issues, and with the training of managers to use design effectively.

The place that design occupies in organizations is a useful starting point for the examination of the role of design management. By examining the traditional classification of design disciplines we can identify how design relates to the various corporate functions, and attempt to show how the value of design can be quantified.

It will never be possible to measure exactly the worth of design in financial terms, but that is true of many of the resources on which managers draw. However, design has always suffered from too much subjectivity; from a view of it as a 'soft' activity linked to the creative and the unquantifiable. Anything that we can do to point out where to look in order to attempt some measurements will help to ensure that managers use design more effectively, and above all treat it seriously.

CLASSIFYING DESIGN

In 1976 I started the design reclassification[2] debate with the suggestion that traditional classifications of designers, based on their professional training, were not always relevant to the needs of the industrial world which employed them. Industrial designers, design engineers, graphic, interior, textile, furniture designers and many others were specialists whose skills sometimes fitted the corporation's needs, but sometimes did not.

If industrialists thought seriously about design at all they thought of it in terms of the products they made and sold, the environments in which they worked, and the information systems they used to communicate their purposes. Thus while a graphic designer or typographer would be the professional used to design a calendar, the considerations and conflicts which arose in the design process would depend on the business use of that calendar. For in a calendar factory a calendar is a piece of product design, on an office wall it is a piece of environmental design and used in a planning department it is a piece of environmental design. Designing for these three uses may well create conflicts, easily resolved for a calendar, less easily, for example, in a major property development, or a mass-produced pharmaceutical product. However, once classified this way the problems are at least identifiable, and it becomes a design management task to resolve them.

This three-way classification is now generally accepted. However, it is often misunderstood or distorted in use. What follows therefore is a restatement of each of the three areas which suggest where we might look in order to measure design performance in each of them. It also identifies some key corporate issues to which design can contribute. With this achieved we can then move on to discuss how design management should be organized for effective use.

Product design

Product design has traditionally been held to be the concern of manufacturing industry. That this is too narrow a view has been clearly understood by the tax authorities who levy value added tax (VAT) at every stage of the product's journey from the origination of the raw materials which form part of it, until it reaches the customer.

Manufacturing industry may be at the centre of this process; but the design of its products is influenced by many other considerations and activities. What all these influences have in common is the extent to which (by affecting its design) they add value to the product, and so contribute to its gross margin performance.

Gross margin is calculated by taking the sales revenue of the business minus the cost of sales after adjustment for the cost of stock. It is the critical and key performance measure of all businesses which deal with products, whether they convert them, make them or trade in them. In fact product design might be best described as 'designing for gross margin by the extractive manufacturing, distribution and retailing industries'. This is particularly easy to see in British consumer products, the design of which is heavily influenced by our powerful multiple retailers. All serious multiple retailers try to become involved in specifying and so designing the manufactured products which they buy in order to sell.

Design affects gross margin performance through its contribution to a range of critical management issues which determine the nature and so the profitability of the product. The list of these critical issues is endless, and they vary from industry to industry, and from product to product. Here are just three of them: *(i)* product innovation in which design is the determinant in the amount of innovation and its rate of flow into the business; *(ii)* quality which is best controlled by designing it in rather than inspecting it out; and *(iii)* product range development where design has a key role in co-ordinating, simplifying and so promoting a product range. These issues, many others of which have been extensively discussed elsewhere,[3] demonstrate how critical design is to gross margin performance.

Environmental design

Over the last decade environmental design has increasingly been interpreted as the province of the architects, interior designers and shopfitters. With the boom in office design, particularly in making use of information technology, and the understanding of the value of shop design by efficient retailers, this pre-emption is not surprising. However, environmental design covers a much wider field, which once again is best described in the terms by which it can be quantified.

Perhaps the most important ratio used by organizations to measure their performance is return on capital employed. All businesses, whether they deliver a product or a service, use this ratio, and all of them recognize that the ratio can be improved in two ways. The first is to raise the return or profitability for a given employment of capital. The second is to reduce that employment of capital for a given level of profitability.

It is in the second area that design has its impact; in helping to determine how and in what the business invests its assets and manages them thereafter. Not all of its assets: investment in intangible assets is not the concern of design. Nor is design directly concerned with the amount of investment in stock (although we have discussed its intimate concern with the nature of that stock and so indirectly in its investment value). Design is mainly concerned with what normally appears on the balance sheet as fixed assets.

These fixed assets comprise buildings of all kinds, factories, offices and shops. The equipment of those buildings, machinery, furnishings, communication equipment, transport and indeed any artefact in which the corporation invests to help it perform its tasks, from an oil pipeline to a pencil sharpener. No sensible business would make significant investment in fixed assets without at least trying to have a say in their design. This is particularly the case in major process industries like petrochemicals, or in service industries like airlines and railways. It is less possible when the investing company is small and the supplier large. However, even a one-man delivery business can paint the outside of a van to his own design.

Environmental design is thus better and more widely described as the 'design of the investment in fixed assets'. As such it has high value in the eyes of management.

Information design

The value of the design of the information systems through which a corporation conveys its purposes to the various audiences with which it is concerned has long been appreciated. This is particularly the case for advertising, sales promotion and public relations materials aimed at external audiences. There is also an increasing awareness of the value of effectively designing for internal audiences, for managers, employees and owners.

The cost of this kind of design is shown as an expense on the profit and loss account, lying between the gross profit and net profit figures. It is usually evaluated as a percentage of sales. Ironically, although the benefits of this kind of design can often be quantified, its costs are trivial in terms not only of those benefits, but also in terms of other levels of expenditure of the kind described above. It is not surprising, therefore, that sophisticated resources exist to supply this kind of design, mainly through the marketing function of the business.

Corporate identity design

Even more sophisticated, but far less well-developed, is the field of corporate identity design. Springing from the work of information designers it embraces and shapes all the aspects of design described above. With easily identifiable, and relatively trivial costs its benefits are more difficult to assess; ahtough some work has been done in the USA to measure the share price of major corporations which have undergone corporate identity programmes.[4]

Corporate identity design is intimately linked to corporate strategy. Because corporate strategy is a difficult area in which to measure success, little attempt has been made to do more than make qualitative judgements of the benefits of corporate identity programmes.

HOW DESIGN IS MANAGED

The four aspects of design management described above are handled by different people with different sets of criteria. It would seem obvious that they cannot be managed in the same way. Let us look at them one by one.

Starting in reverse order with the management of identity, it is invariably the case that whatever the organizational shape of the business, whether it is centralized or decentralized, functionally or geographically deployed, corporate identity design must always be a central resource.[5] Only in this way can it comprehensively influence and modulate the various design activities of the business. However, each of these activities (as described above) needs to be managed in a way which is relevant to the way in which the business itself is organized. Thus, if gross margin accountability is delegated to divisions then so must be product design.

Furthermore, because product design is operational in style and directly relevant to performance, it is likely that line managers will be deeply involved in the design management of products, though they may not be aware that they are so involved. Recent research at the London Business School has shown this to be the case.[6]

Information design is also intensely operational, particularly when it is closely linked to product, through packaging and advertising. When this is the case, it too follows the organizational pattern of the business, usually through the hands of brand managers, buyers, and similar management roles. When information design is more intermittent it is usually managed closer to the top and through a staff function. The design of corporate advertising, public relations messages, management information systems and data recording systems are typical examples.

The design management of the environment is characterized by abrupt shifts from continuity to discontinuity. The design of major investments is invariably done by outside consultants who are used even in the larger corporations. However, the maintenance of such invest-ments is often in the hands of facilities managers,[7] a fast-growing profession of design-trained people. Line managers are much less likely to be involved in managing environmental design except where the corporation operates in discrete but continuous environments. Obvious examples are multiple retailers, banks and brewers, restauranteurs, hoteliers, and a range of organizations in the communications business.

ORGANIZING FOR DESIGN

It will now be apparent to the reader that it is far easier to determine what design management is, and how it is valued, than to lay down rules on how it should be organized, and who should do it. For identity management the answer is clear. Other kinds of design management will (indeed should, if they are to be accountable) be organized in ways which relate closely to how the rest of the corporation is managed. Any large complex corporation is likely to manage its design activities in a number of ways and at different places in the organization with differing reporting responsibilities. So the much-discussed issue as to whether design should be represented on the board or whether it is managed as a staff or line activity begs the question.

Should this worry those of us who proselytize for design and its management? No more than it worries those who support the cause of finance, personnel, computing and the other functional management activities. Indeed what the design profession can learn from them is the flexibility that they offer their professionals, and the freedom with which they provide others with access to their skills.

Take computing. The profession of computing science has gained enormously in influence since it came out of the closet. Once a black-box mystery locked away in air-conditioned rooms and operated by mysterious boffins in white coats, it is now accessible to every line manager who learns about it and uses it as an essential tool. Furthermore, this is not one-way traffic. People trained in computing, like accountants, personnel people and marketing people before them, have themselves moved over to line management, especially in industries where their skills are relevant to the business purpose.

This dismantling of professional frontiers still has a long way to go in the design world. The key to it is training. Once managers learn to appreciate the value of design, then deploying design resources for effective use (our definition of design management) can be left in their capable hands. The task is likely to be made easier by their recognition that, albeit in a disorganized way, they are probably already managing design.

Earlier on in this paper I indicated that some London Business School research was revealing that this is indeed the case. Indeed, many managers are 'designing' and managing design without recognizing that they are doing so. Dumas has called this phenomenon 'Silent Design', and it provides the firmest possible base on which to build formal design management training. That the base already exists is not, on reflection, that remarkable. After all, everything that is made has to be designed and the process consciously or unconsciously is likely to have got itself managed one way or another. Our task, through training, is to ensure that it is done well.

REFERENCES

1. Gorb, P. 'Design and its Use to Managers', *Royal Society of Arts Journal*, November 1979.
2. Gorb. P. (ed. and introduction), *Living by Design* (Pentagram), Lund Humphries, 1978.
3. Lorenz, C. *The Design Dimension*, Blackwell, 1986.
4. *Wall Street Journal*, textline reference 10 July 1987.
5. Olins, W. *Guide to Corporate Identity*, Design Council, 1984.
6. Gorb, P. and A. Dumas, 'Silent Design', *Design Studies*, Vol 8 No 3, July 1987.
7. Duffy, F. 'Intelligent Office Buildings', *Facilities*, Vol 5 No 3, March 1987.

PART I

VIEWPOINTS

James Pilditch founded one
of Britain's leading and
largest design consultancies.
In recent years he has been
deeply involved in directing
a number of Government
activities which seek to
promote design in Britain.
He is a member of the
council of the Royal Society
of Arts and has published
widely in the field of design.

Using Design Effectively

In Britain today, thanks to the vision and drive of the Government, the Design International Analysis, the pioneering work at the London Business School and to many others – in education for instance, where impressive things are happening – design is surging like a wave that cascades over every boat and jetty in its path.

It would be a deaf and blind man who had not heard about design in the last few years. The thing to do, harassed businessmen are told time after time, is to use design, bring in a designer. Well, you will not catch me arguing with that. However, there are snags. Offering design as the latest star to follow can promise false hope, unless, like a precocious child, it is kept in balance, in its place.

I am thinking of the Pope on the balcony on Easter Sunday, raising his hand. Thousands crowded into the square beneath him will be blessed as he does so. I say this because too often design is sanctified, believed to be inherently good. Take some . . . you will feel better. Sadly, by and large, designers do not enjoy that omnipotence. Like dentists, doctors and deep-sea divers, designers come in all sizes; some marvellous, some workaday. The best designers help shape our world, enhancing life, enhancing business. Not all reach these heights. Nor, of course, does everyone who employs them.

As a design consultant I have always said: 'See good design and you see a good client'. Actually, clients, people who point the direction and take the decisions, get too little credit. However, for every wonderful thing we see, there are others; not effective, not value for money, not even pleasing. So some clients, like some designers, are better than others.

Fortune attributed Ford's renaissance in the USA very largely to its design drive. As that suggests, the best uses of design can be marvellous – turning around whole industries. However, it is as true to say that the misuse of the design process can be as daft as the opposite is wise.

At the Royal Windsor Horse Show you will probably see a tent where they sell garden gnomes. Not any garden gnomes, but gnomes postured in quite un-elflike or Enid Blyton-like ways. With their drawers down. These gnomes were created in much the same way as the Olivetti PC, the Braun shaver, the Saarinen chair, the Ford Sierra, or whatever you think is a designed object.

In all cases, someone had to have an idea of what their customers want. They had to decide what features to offer, what form the object should take, how to make it, how many to make, what it should cost, and all the rest. A design had to be drawn, a model made, mechanical drawings done, prototypes tested.

The garden gnome example may not be a solemn enough example, but what I am saying is true of every door that does not fit, every shirt that shrinks, every poor product that deservedly does not sell. Everything in our man-made world is designed by someone; but only too often carelessly and unthinkingly. In short, although design is not an option, whether to use design effectively and with intelligence is, indeed, an option. The good, the bad, the ugly, all are designed.

Let me turn to the ugly. People think that design is about aesthetics, creating objects or spaces or whatever with good form, sensitive colours; good manners, you might say. Though the attitude may be weakening, designers have been seen as the people to add taste or style – after all the serious work of planning and creating something had been done; like the chrome trim they used to fit on cars. Engineers are particularly prone to this opinion. They are the true designers they say. Others, as one eminent engineer put it to me, are mere 'paint-brush boys'.

I once spoke in Geneva to what might have been called 'The Executives Club'. Businessmen in the audience could not understand what design had to do with them. 'It is for living rooms, is it not?' one asked. Curtains and carpets and chairs; stuff like that. The decorative (and that means unimportant) things in life, they implied.

I told them about a bread-slicing machine. One day a soldier went to see the MO. He had lost a finger. Within a few minutes, another soldier arrived. He, too, had lost a finger. How had it happened? 'Well,' said the second man, 'my mate and I work in the cookhouse. He cut his finger off in the bread-slicing machine and I was showing the others how he did it.' Who could call that machine well-designed, whatever it looked like?

Or think about the late and much loved Sir Misha Black. During the war he worked for the Ministry of Information, in the basement of the National Gallery. 'What did you design for them?', I asked. 'Horse droppings', he replied. Horse droppings? Design? Using design effectively? The story was this. Resistance workers in France needed an explosive device they could lay on a road quickly to blow up enemy trucks. But it must not be noticed. The brilliant answer, in those days, was horse droppings. You would not spot them. Try my earlier theory again: someone, perhaps Misha, had to say how big the item (I do not know what to call it) should be, what it should be made of, how to fit an explosive device into it. He had to make working drawings, test prototypes, and the rest. Was that an effective use of design?

Do not get me wrong. I am not denying good form and good looks. I am struggling to say that effective design combines analysis and imagination, practicality and sensibility, all the arts you need to make new things work. And that includes selling in sufficient numbers to employ and generate wealth.

Contrariwise, I assert that good form, appropriate use of colours and textures and finishes, whatever you think comprises something that is visually and sensually satisfying, are themselves often instruments of effectiveness.

Every designer I know carries a torch, sometimes concealed, sometimes not, to create beautiful things. God knows, the world could do with more. But, while honouring this, I do think it only ethical for them to keep their eye on the ball the client pays them to look at. First things first. Happily, the poles – of practical effectiveness and visual delight – need not be poles at all. Anyone who has seen Tom Watson hit a golf ball will know what I mean. The best things both work well and look right.

What design should almost never be used for is as a kind of fancy wallpaper to cover cracks. If you have a dud product, spray a little design on it. Years ago, I remember, a man in a drug company talked about getting a new corporate identity. He wanted, he said, 'an image of low white buildings on spacious green lawns'. His dingy office, I should add, nestled behind Euston Station; held up, as far as I could see, by centuries of soot and not much else. Rather rudely I said what he should do is 'buy a field and hire a good architect'.

In other words, truthfulness and integrity must be essentials of good design. Laying aside any moral consideration, I am suggesting that if the design is not true it will not work for long. Idi Amin – remember Idi Amin? – he gave himself the VC. But that did not confer extra courage on him. Saying that, it is certainly possible to reinforce and accelerate acceptance of your truths. Think of the way 3i's consistently splendid graphics have softened the marble-halled dourness and dullness of banking. They show for all to see that 3i is more human, more lively, more approachable.

However, that message, to be effective, must be not only credible but relevant and that obviously is the case with products. If you do not provide what people want you will not sell much. However, it is true of other kinds of design too. I am thinking of a bookshop whose sales fell after it had been designed. From being a lovely, browsing, muddly, caring-about-books kind of place, the designer made it cool and grey and sparse and tidy. For Heaven's sake, what book-lover wants tidy? In other words, appropriateness is the key.

I remember once writing about Jack Dash, the dockyard union leader. He was punched one day by a docker's wife. She felt like punching him and punched him. Now, with the exception of members of the MCC one infamous day (and that was so exceptional it made the front page of *The Times*), most people do not punch others. They may think 'a little of what you fancy does you good', but most of the time their emotions are under control. There are still others, of course, who are 'only rude intentionally', so controlled are they.

There are important conclusions from the punching episode which are relevant to my argument. First, people vary, and part of their variety is their emotional response. Second, we know that emotional drives are quite as influential as any other drive. Third, design is a route to the emotions or it is nothing. Fourth, the key to successful marketing, we would all agree, is segmentation and that includes emotional or psychological segmentation.

The global notion of mass marketing is too simple. 'If you're not thinking segments' said Theodore Levitt, now the editor of the *Harvard Business Review*, 'you're not thinking'. We know that, and slot people neatly into their so-called socio-economic group. But that is as clumsy as categorizing cricketers by their collar size.

I have always argued for what I have called 'psychological segmentation'. I mean identifying the whole nature of people, not just the size of their house, but satisfying their wants as well as their needs.

To be effective, design must be appropriate to its audience, bearing in mind that, increasingly, people have the freedom and choice to satisfy their whims, in their own way. We must appeal to the illogical as well as the enlightenment in all of us; to people with passions as well as paychecks – to people as they are. I think the car companies understand this, and perhaps the airlines. But how many other industries do? Let me turn to a simpler point.

I worked as a consultant to a biscuit firm years ago. 'Do they test their products?', I asked. 'Yes', the chairman said confidently. Every week they held a tasting panel. He was there, the chief baker, this one, that one. All experts. 'What about the customer?' No, she was never asked. 'We know what makes a good biscuit', he said, 'customers don't'. What happens if the shopper wants more cream, or less toasting that the experts know is right? Is she to be dissatisfied; or is the company, by not selling its goods? In other words, to design effectively we must reconcile our sometimes informed judgements with those folk who may be less-informed; our customers. Although a truism, this is a complicated, even a moral, issue.

This leads naturally to finding out what our customers want. As-toundingly, according to the Institute of Marketing, no more than three out of ten firms in Britain do any market research at all. How, then, can the client or designer know what his customers want?

There are only three possible answers. One is that he assumes he knows. Indeed, that was the reason most frequently given to the Institute of Marketing by people who used to market research: 'We know enough already'. Another is that he guesses. A third, to quote the immortal words of Rhett Butler in *Gone With The Wind*, is that he 'doesn't give a damn'. In other words, we impose on the market what we believe to be right.

As markets spread and fracture, as our competitors strive to come ever closer to their customers, none of those answers work. Design will rarely be effective in any of those ways – or not for long.

Nevertheless, I absolutely acknowledge and rejoice that sometimes good people, including designers, can intuitively feel what's coming, what's wanted. When there was a Swan & Edgar, a Canadian designer in my office, Neil Shakery, did some work for them. To my horror, he designed a shopping bag that was blue and green. Now, as a student I was taught the maxim 'blue and green should never be seen'. But there it was. Within eighteen months every secretary in London wore a green blouse and blue skirt, or the other way around. How did the designer sense that?

I have no idea how such intuition works. But I do know that to exclude it, because it leads to the unexpected, is nonsense. To get the best from design and designers, clients sometimes have to accept the inexplicable. However, allowing for that over-riding, even extrasensory intuition, I am convinced that effective design flows from knowing who you are aiming at and what it is they want.

This knowledge is not just left to the accredited researchers to find out. At Honda and Sharp and elsewhere, designers are away from their drawing boards for up to half the year – talking to customers in shopping malls, visiting dealers and so on. At 3M everyone in the R&D department makes sales calls. The president of one US company makes it a firm rule to phone six people every week: three who bought his company's products, three who did not.

Which firms, you might ask, are more likely to use design effectively: those that 'know enough about their customers already' and say to their designers 'you're here to work, where do you think you're going?' or those that are out and about – listening and watching and talking to people they aim to reach?

Sir Misha Black once said a designer should be a 'jack-of-all-trades and master of one'. In the same way, to compete effectively, a product needs to be good all round, and better than any competitor in the one dimension that matters most to customers.

If you buy a washing machine what matters most to you? Is it the speed of the wash? The number of revolutions per minute? The quietness of the machine or its reliability or appearance of the price? Which? All must be good, of course, but which is the really telling point to you? The answer is that different people will give different answers. They will vary by customer, by region, all sorts of things. What I am saying is that to be effective design must help you be better where betterness counts.

In Seiko, where they produce a watch every six seconds, I was staggered to learn that they can and do make as few as fifty at a time to suit particular customer groups. Then, they switch in half a minute to another batch. At Ricoh, the world's largest maker of copying machines, they do the same. Now that is what you call segmentation. That is what you call targetting and tailoring. That is what I call using design effectively.

Andrew Carnegie once said 'There's no limit to what you can achieve if you get up early enough and stop smoking'. Focus design well, and work hard, and you can indeed, create goods (or other things) faster, cheaper, easier to use, more attractive, more energy-conserving – whatever dimension you choose, but that effort (and stopping smoking) will be wasted if the dimension you improve is not the one that matters most to the customer.

I shall labour this point because I do not think we usually look at design this way. It is the totality we judge, not the critical detail; the sword we look at, not its cutting edge.

Plessey once developed a piezoelectric crystal gadget to light the gas. Made to their habitual defence contract standards, it was beautifully engineered. However, it cost over £9 to make. Was ruggedness and reliability the key to competitiveness? Most unlikely. At their request, David Crisp designed a plastic thing that clipped together and sold for, yes sold for, £1.75. It lasted too. At home we still use one daily now, after ten years or more.

If you are designing a jet engine, however, or cable car, the story would be different. You saw when the Boeing 747 was battling it out with the Concorde how low operating costs and seat-miles mattered more to operators than high speed. Design, I shall say it again, is not absolute. To be effective it must be relevant to the customer's perceived need.

This says that from a whole quiver full of reasons for purchase, you must pick the one that counts most to the customer and go for it. You will not win by designing something that is good all round and not better than any competitor in the one essential dimension. Oh, you might do well enough, but not strike forward, in front. That tells us that while being 'obsessive' about customers, as Tom Peters and Bob Waterman put it, we have to be more than keenly aware of what our competitors are up to.

It is competitors who set the standards we must beat. It follows that to use design effectively – and for the moment I am taking that to mean creating something better that people prefer to buy – you must know all the choices open to the customer. So knowledge about competitors is vital to effective design. I do not mean what you pick up at trade shows, but real, detailed information. There is much more to say about that.

Let me turn to an equally important issue. Just as using design well can be inhibited by not knowing enough about our customers or our competition, so, of course, may we be our own barriers.

A good story first, then a poor one. My firm was employed to redesign a rivet-setting machine, a great cast job with more protruberances than you would find on Chartres Cathedral. The new design was radical; a steel stand, all the working parts clad by a simple box. The new machine, or range of them, cost something like half to make. It was safer to use, simpler to understand, better. The managing director of the firm said 'we could not have made that, you know, because we know how to make rivet-setting machines'.

Contrast that with this experience. Dave Muston redesigned some ear defenders. The old ones had maybe fifteen or twenty parts; metal soldering, screw fittings and so on. David's design consisted of three plastic bits that clipped together. It was rejected. The sad sequel was that within a couple of years that firm was driven out of the ear-defender business by a competitor. His product? Three plastic bits that clipped together.

Perhaps the important lesson from these examples is to tell designers where to aim, not how to get there. So often freshness and imagination, the courage to dare, are stifled by imposing, preconceived (and mostly old) ideas. The great thing is to use designers for what they are good at. Their restless discontent, their 'must be a better way' mentality, is to be grasped and cherished, not cut down to size.

What is good about good designers is not that they draw well (many do not, as a matter of fact) but that they think well; differently, freshly, often with the incisiveness of ignorance and innocence. I mean that as a profound compliment to them. To use design effectively you have to create the climate in which ideas can flourish. The ordered, disciplined corporate structure is not ideal. Nor is Harold Green's famous 'no surprises' attitude.

Creativity is not a tidy process. Do not be disheartened by disorderliness. Even the much-admired 'scientific method' is not that orderly, I suspect. It gets cleaned up when the thesis is written. Archimedes had his best idea in the bath. And Darwin could point to the exact spot in the road where, out in his dog-cart, the key to his theory of evolution struck him. Not much to do with nine-to-five. Creativity is not neat at all.

To get the best, you have to confide in people your hopes and plans.

And show you care. Caring counts more than counting, and this is especially true with creative people. In the 'winning' companies I wrote about it is common to see people on the main board drop by the design department every month. Design, indeed, is invariably at the heart of the business, not stuck in the bowels, like a pretty canary down a very black coalmine.

One way in which counting can help is when you see first ideas. It is highly important to generate many ideas, because it is easy for us to fall in love with our first idea. Unless we are shoved, we never budge from it. Instead, we polish the presentation until it shines. That way creativity does not lie. It is better, in my experience, to see rough work, to say that is great. Let us keep that, but have you tried . . . ?

The tone needs to be positive. Perhaps like everyone, only more so, creative people of any kind – and I include anyone trying to find new ways – need lots of encouragement. Endless stroking, tireless support, lots of praise, lots of recognition.

Nor should the recognition be only or even financial. I am working with a firm now. What their people want, their kind of recognition, is to be sent to a business school. In other companies the 'pin-ball' idea works. If you do well, you get to play the game again, get selected for the next project. Identifying people publicly, as firms almost never do, is also important. Three good people joined my firm because they had seen we always credited designers with their work. So recognition of the people who actually do the work is as important as, in my opinion, it is just.

Speed is important. It is increasingly true that nowadays you can have a design, beautiful in every way, which will not get you anywhere unless you are out with it quickly. How quick is quick? Sony allow six months from having an idea to having the product in the market. That quick. When Brother introduced the first electronic portable typewriter they had a lead of two years. Now, when they launch a new product they reckon to have a lead of four weeks. That quick. The model life of a fax machine before something newer comes in, is put at sixteen weeks. That quick.

Here is another example of speed in design. It comes from Akio

Morita's excellent book *Made in Japan*. Yamaha decided to challenge Honda for a larger share of the US market for small motorbikes. Honda responded at once. How? According to Morita, they launched a new model every week for over a year. That kept Yamaha out of their kimonos. Hearing that, would you not say our departmentalized, systemized, sequential way of going about things does not have a hope?

How to use design effectively is to get a move on. Halve the time you take to develop a new product or any other kind of design. Others have, we must too. You must find out how to do that, and your designers must find out.

One answer, a key one, is teamwork. The essential point is that it is not designers alone who design things. It is designers working with manufacturing, with marketing, with finance and many others. Angela Dumas, at the London Business School, has identified what she has called the 'silent designers' – people in the firm who affect design decisions whether they know it or not. Olivetti have 'customer service' on their design team. Others bring in suppliers. Sony have social psychologists alongside designers. We must beware of the department and division heads that protect their patch. As always, the Americans have a word for this. They call it 'turfiness' – guarding your turf. That must go if we are to get anywhere.

To use design effectively – and that means pointedly, comprehensively and swiftly – all the right people must be together all the time. Tom Peters and Nancy Austin, who wrote *A Passion for Excellence*, coined the phrase 'Try it. Fix it. Try it'. What they say is: test real products with real people as quickly as possible.

In Japan we asked Sony about their speed of product introduction (from idea to market in six months, remember). Is that not risky? Does that not lead to mistakes? Yes, they replied, but were unrepentant. 'You will learn far more in the market in a couple of months', Sony said, 'than ever you will in the laboratory in a year'.

If we are talking about using design effectively, Sony is not a bad note on which to end.

Angela Dumas **3**

Allan Whitfield *Allan Whitfield is Head of the*
Design Department at
Teesside Polytechnic. His
primary research interest is in
attitudinal modelling, though
he has published extensively
in the fields of aesthetics and
colour science. Angela
Dumas is Research Fellow in
the Design Management Unit
at the London Business
School. She has recently been
appointed Design
Management Consultant to
British Rail.

Why Design is Difficult to Manage

Over the past decade design has emerged as a potent economic force in both the manufacturing and service sectors of Western industry. The notion of technology led is now being replaced by technology-design led. The emergence of design, however, has been problematic for industry. Technological developments are relatively clear-cut, design developments are not. The formulation of effective design policy and the management of design are fraught with difficulties for companies. The research reported in this paper sought to investigate current practice and attitudes towards the management of design in British industry. This involved a survey of senior managers from major companies in Britain. A similar survey has been carried out in the USA. The main findings to emerge were: *(a)* the existence of four distinct types of company, each with their own approach to design; *(b)* the pronounced effect a design manager has upon attitudes within a company; and *(c)* the clear distinction between the manufacturing and service sectors. The practical implications of the research are discussed here with reference to the management of design in companies.

THE ORGANIZATION OF DESIGN MANAGEMENT

Recent government initiatives and the efforts of key people in industry and education have raised the profile of design and its importance to industry.[1, 2] These initiatives have led to an increasing number of companies seeking to adopt and implement a design policy. However, it has become clear that if industry is to take full advantage of the design opportunities that exist, then they need practical guidelines to help them utilize design effectively and efficiently. Managers need to know more about what design is and how it can be employed in a particular company. They need to understand more about the design process and its organization within industry, and to be able to separate this from the common preconception of design as a 'creative' activity carried out by others. In particular, attention needs to be given to the organizational structure of the company if the interactive nature of the design process is to be managed effectively. Also, the participants in the design process need to be identified and their contributions co-ordinated.

A research programme based at the London Business School in the Design Management Unit has been looking at how British industry organizes for design. The study set out to establish what is current managerial practice and what attitudes are held about design by senior managers in both the manufacturing and service industries. The purpose of the research, therefore, was to gain a better understanding of how companies actually use design as a part of their business. As the word 'design' is used in so many ways a definition was adopted to describe the concerns of the research. The definition was 'a course of action for the development of an artefact or a system of artefacts'. This encompasses both the aesthetic and technical aspects of artefact development, and covers activities commonly associated with both designers and design engineers.

The research programme

The research was conceived in four phases: *(a)* a pilot study to establish focus; *(b)* the development and piloting of a questionnaire; *(c)* a company study to examine the design process in depth; *(d)* administering the questionnaire to companies. This article focuses upon the pilot study and questionnaire.

The pilot study Twenty companies were approached to participate in the pilot study, ten from manufacturing and ten from the service sector. Two of the key criteria for selection were a declared commitment to design and a formal design policy. All of the companies selected fulfilled the 'commitment to design' requirement and approximately two-thirds the 'formal design policy' requirement. Of the twenty companies approached, sixteen agreed to participate.

Informal interviews were conducted with three to six people in each organization. An initial interview with the chairman or chief executive provided the first contact and thereafter a programme of interviews was arranged. Co-operation was good.

Following the interviews, matrices were developed to map different aspects of each company's use of design. The matrices provided a comprehensive and inclusive overview of the design process; in addition, they allowed activity in the design process to be recorded individually for each company.[3]

During this initial stage of the research it became clear that some individuals were unwittingly making contributions to the design proces; that is, they were participating in design tasks but attributing these activities to other tasks within their own functional area. There was little acknowledgement or apparent awareness of the design process as a continuum despite the fact that many individuals were actually participating in it. Their participation was frequently sequential and embedded within other tasks normally associated with their own functional areas. This led to problems of communication, and variations in their description of the tasks, usually phrased in the language appropriate to an individual's function. Individuals generally failed to perceive the task as a subset of a process. Maximizing potential relies on good communication and shared goals; both of these seemed to be largely absent in relationship to design.

The pilot study concluded that managers were *(a)* active in design tasks; *(b)* contributed to the design process; and *(c)* frequently worked alongside professionally trained designers. However, acknowledgement of their own contribution was, in general, absent. The phrase 'Silent Design' gives a name to this situation. A paper with this title was written to describe the method, outcomes of the pilot study and an agenda for subsequent work.[4] The pilot study also identified the following as areas for detailed investigation within a questionnaire.

- Managers' perceptions of design tasks.
- Functions and job titles of those managers involved in design tasks.
- The role of the designer and the design manager.
- Financial accountability during the design process.
- The organizational structure in relationship to both discrete design tasks and the total design process.

The questionnaire survey In designing a questionnaire for industry the decision was taken to adopt a broad rather than a narrow focus. This was justified on the grounds that, surprisingly, no previous empirical investigations had taken place in this area and, consequently, no data base existed from which a more precise focus could be obtained. The decision to take a broad overview meant that the questionnaire was relatively large, taking approximately one hour to complete. In view of this and the nature of the subject the questionnaire was designed to be completed in a personal interview.

The questionnaire consists of five sections. Section One identifies an individual's job title and his or her tasks, further education and experience. Section Two covers general company information and the financial and organizational aspects of design. Section Three focuses upon design practice within the company. Section Four deals with the financing and organization of design within the company, with Section Five focusing upon its management.

While it was intended to produce data which would allow comparisons between different industry sectors, it was considered equally important at this stage to gather information across industry sectors about specific issues, such as accountability for design, the use of consultant and/or in-house design teams, and authority and control of design activities. The opinions and attitudes of individuals within functions were also important issues. Information was sought on how individuals perceived the whole area of design and its organization within the company.

The questionnaire was targeted at middle and senior management from each of the following functional areas: marketing, finance, production, operations and design. It was administered to seventy-six people, between three and six in each company, covering both the service and manufacturing sectors. The manufacturing companies were those producing either consumer products or products where design was concerned with both engineering and appearance.

Results The results are unequivocal in indicating the high importance attached to design by the participating companies. The results are equivocal, however, regarding what constitutes design, who is responsible for design, whose budget finances design, the availability and coverage of design guidelines, and the power of design managers. In short, those interviewed acknowledge the need to manage the design process effectively; however, they appear to experience difficulty in identifying the design process and accurately positioning it within their respective companies. Overall, the design function is considered to be of strategic importance; but, since it is unlikely that those involved recognize the activities of all participants in the design process, it is also unlikely that a structure exists which would enable people to work together effectively. Design, as a consequence, emerges as diffuse, lacking in definition and difficult to use as a strategic tool.

There are two factors that appear important in creating this sense of diffusion and lack of definition. The first revolves around the extent and nature of the design function. Here, not only is the term 'designer' used to refer to a number of professions, with engineers as a prime example, but there also emerges the problem of the 'silent designers' referred to earlier in this paper. An example of the engineer in the role of designer comes from the manufacturing sector. A senior engineer who had responsibility for the design department identified himself as the design manager and answered throughout the interview in this role. However, all of the other senior managers from this company who participated in the survey felt unable to answer questions on the role of the design manager because they did not consider that their company had a design manager. This highlights the confusion that can surround design: the 'Silent Designers' who unknowingly participate in design tasks are scattered throughout organizations. A further example from the service sector, a retail company, was a manager who took a wide range of design decisions on the packaging of merchandise including size, shape, suitability for customer handling, display arrangements and suitable materials. He considered that he had nothing to do with design decisions and, therefore, introduced the interviewer to the company's professionally trained graphic designer. Her sole contribution to the packaging design was, in her words, 'the picture on the box'. As long as managers can only identify a fragment of the design process as design, the sense of diffusion and lack of definition will remain.

The second factor contributing to this situation is that industry sectors differ in their approach to and attitudes towards design, particularly in finance and control. For example, the service sector considered that their budget problems were frequently caused by poor information from designers, and that tight budgetary control was needed from the top. The manufacturing sector did not share this view. The service sector's concerns with budget and information about design should be seen in the context of its particular products. For many these may be the high street retail outlets which sell a service, as do the banks, or a product which they may package but not manufacture, as do the multiple stationers, chemists and supermarkets. Regarding retail outlets, the opportunity to market test the selling environment is extremely limited. Even the multiple store conversion is, to a large extent, a series of one-off production problems, despite the standardization of some components such as display systems. In this situation good communications and up-to-date information will be crucial. Without these, budget problems will occur. Similarly, the absence of a structure to accommodate the total design process will contribute to communication breakdowns. In these circumstances tight control from the top would appear the safest answer. However, addressing only fragments of a design process can lead to over-control of the identified parts as a compensation for lack of control over the unidentified parts. Logically, this leads to an unbalanced structure where measures of effectiveness become more and more sophisticated over time but cover only a fragment of what really needs to be measured – a problem identified by Tichy[5] when managers confront change. This can lead to confusion which is exacerbated by the differences in general management, financing and control of proejcts, including design, between the manufacturing and service sectors and between companies within each sector. These differences are typified by the role of the design manager. Only half of the companies participating in the study had a manager solely responsible for design, though this person's sphere of influence was very varied. Of those with a design manager 23 per cent had board representation. However, this must be set against a background of non-designers generally retaining control of the design function at board level. These are predominantly the marketing function (63 per cent) and engineering (42 per cent). Those companies without a design manager had no design representation at board level. The situation is not unlike a set of principalities, each with a different language, culture and attitudes.

The study identifies areas of strong agreement – largely the importance of design to the companies and the need to effectively exploit it – and evidence of considerable divergence – who is accountable for design, how is it organized and who participates? The question emerges, however: are these differences systematic? In effect, is there order within this diversity? To answer this question a multidimensional scalogram analysis (MSA) was carried out on a subset of the questionnaire data. This analysis enabled common features and types to be identified.

The results of the analysis reveal the existence of two main, orthogonal divisions which account for much of the variation: *(a)* the existence or not of a design manager and *(b)* the type of company – service or manufacture.

Is there a design manager?

The existence of a design manager exerts a considerable influence upon attitudes and practices within the company. This is characterized by the beliefs of those involved in the study that *(a)* design managers should have power and *(b)* a high level of accountability exists for design. In addition, where a design manager exists within the company, design projects are more likely to be structured similarly to other projects and considered not to require central control. Other effects can be noted from a related analysis. Those with a design manager favour design operating as a profit centre and perceive the design function as exerting influence upon company policy in staffing, finance and project development. They also consider design to have greater importance within the company. Those without a design manager do not share these views.

Type of company

The second major influence is determined by sector differences. Service companies are characterized by their use of design consultants, the existence of design policy documentation and also the domination of the design function by marketing. Manufacturing companies, in comparison, are characterized by their use of internal design teams, a general absence of design policy documentation and the strong influence upon the design function of engineering. Related analysis reveals that while both sectors consider design to be equally important – in fact, the degree of similarity is remarkable – the services sector views designers as less accountable than the manufacturing sector.

An interesting feature of the above divisions is that the presence of a design manager appears to exert a powerful attitudinal influence within the company, while the status of the company – service or manufacture – is less attitudinally influential than practically (e.g. design documentation; use of internal or consultant designers). As indicated above these divisions are orthogonal. When combined, four major types emerge from the analysis. These types represent four distinct company attitudes and practices towards design (*see* Table 1).

Manufacturing industry with design manager This type is characterized by the domination of the design function by engineering, the structuring of design projects like other projects, and the use of internal design teams. The designer does retain some accountability for design.

Table 1 Typology of attitudes and practices towards design

A Manufacturing/Design Manager	B Service/Design Manager
Engineering largely accountable for design.	Marketing largely accountable for design.
Designer has high accountability.	Designer has high accountability.
Use of internal designers.	Use of external design consultants.
Design should not be centrally controlled.	Design should not be centrally controlled.
Design manager should have power.	Design manager should have power.
General absence of Design policy documentation.	Existence of design policy documentation.
Design projects are structured like others.	Design projects are structured like others.
C Manufacturing/No Design Manager	D Service/No Design Manager
No clear accountability for design.	Marketing largely accountable for design.
Designer has low accountability.	Designer has low accountability.
Use of internal designers.	Use of external design consultants.
Design should be centrally controlled.	Design should be centrally controlled.
Design manager should not have power.	Design manager should not have power.
General absence of design policy documentation.	Existence of design policy documentation.
Design projects are not structured like others.	Design projects are not structured like others.

Service industry with design manager This type shares with the above a similar structuring of design projects and some degree of accountability by the designer. It differs from the above in its reliance upon design consultants and design policy documentation. Similarly, while engineers have some accountability for design, marketing has the dominant role.

Manufacturing industry with no design manager This type is characterized by a low level of overall accountability for design, with no clear focus; for example, neither marketing, design nor engineering appear accountable. Further characteristics are: *(a)* the reliance upon internal designers; *(b)* the belief that design should be centrally controlled; and *(c)* the recognition that design projects are not structured like other projects.

Service industry with no design manager As with service industry with design manager, the focus of accountability for design lies with marketing. Design policy documentation is available, and reliance is placed upon consultant designers. Design projects are not structured like other projects and design is considered to need central control. The designer has low accountability.

The very existence of this typology is interesting. It is also significant for two reasons. First, it reveals the diversity of attitude and practice encountered in industry. Industry, in its relationship with the design function, is not a uniform domain which can be addressed simply and directly; rather, it is segmented with each type exhibiting a unique culture/practice profile. As such, it argues against a simplistic approach to the management of design within industry and suggests a greater need for tailor-made solutions. Second, the existence of such a typology helps to account for the difficulty of addressing questions of design management within industry and also the apparent diffuse nature of design. Design, as it is understood and practised, differs markedly within industry.

Having identified two major divisions and their incorporation into a typology, it should not be assumed that this exhausts the range of either types or influences. A particularly potent attitudinal influence can be observed in the backgrounds of the senior managers involved in the survey. The most pronounced distinction was found between those with a design training and those trained in other professions. The 'designers', as they may be termed, held views that separated them from the other professions (marketing, operations, finance and production) in their perceptions of design within the company. For example, the designers consider that: *(a)* the design function has greater influence with respect to financial, staffing and project expenditure within the company both for design and non-design projects, *(b)* the Design Studio Manager and new Product Development Manager should have greater influence on the allocation of resources, while the Project Manager should have less influence. (It is notable that the collective view of the other professions is that the order of influence should be: *(i)* Marketing Manager, *(ii)* New Product Development Manager, *(iii)* Design Studio Manager, *(iv)* Project Manager, *(v)* Production Manager, *(vi)* Sales Manager.) *(c)* The Design Manager is perceived as having greater power, greater reward and greater frustration. *(d)* The job of designer is seen as carrying greater accountability within the company. Not surprisingly, the designers consider that design should be a central function within the company. They also contend that budget problems are not caused by poor information from the designer, nor costs associated with design projects, nor production costs, nor lack of common goals, nor interference from the top. In all of these views they differ markedly from their colleagues.

It would be reasonable to expect designers to be more 'focused', 'better informed', 'sensitive' to a questionnaire dealing with design management; however, the differences are pronounced. They evidence more than mere enthusiasm for design and its importance within the company; they indicate a radically different view. It appears that the designer's conceptual model is at variance with the shared conceptual model of the other professions. It also suggests that their perception of the company 'reality' is somewhat 'unreal' – measured against the collective reality of the other professions. This begs the question: if their conceptual model is dominated by design will their conceptual model of other aspects of the company, its mission, structure etc be at variance?

Can design managers understand the other's conceptual world with reference to design? Can they communicate with the others effectively?

In contrast with designers other professions exhibit fewer differences in attitudes to design than might be expected. This is particularly noticeable in manufacturing industries where both marketing and production people are more similar in attitudes than conventional wisdom would dictate. While the marketing function retains a high level of influence over design the attitude of other managers is generally towards maintaining this role. This may be because it is all they know or they may see advantages in this. There are indications which lead us to think that there are some quite distinct disadvantages in the maintenance of this role, if the design process is to be utilized effectively as a company resource, particularly in the manufacturing industries.

Other notable findings that emerge from the data analysis are:

1. In companies with no design manager the production and project managers are more influential: their influence is less when a design manager is present.
2. If design is a section of company corporate documents then *(a)* design is perceived as needing greater budgetary control, *(b)* the design manager is likely to have greater financial responsibilities for design related projects and *(c)* budgetary control by production and project managers will be less. Whether design is a section of corporate documents or not has little effect upon the influence of the marketing manager: this remains strong.
3. Production and operations managers consider that design needs greater budgetary control. Marketing managers do not agree. This difference may reflect the fact that marketing already maintains a strong budgetary control over design. Alternatively, it may indicate some lack of awareness on marketing's part of the problems of production/operations control associated with projects.
4. Production and operations managers regard the job of design manager essentially as one of developing new products. Marketing managers are less inclined towards this view, possibly seeing it as an encroachment upon their own role. Marketing's view of the

design manager's job is to oversee design, a view shared by the designers.

IMPLICATIONS

What practical implications can be gleaned for the organization of design in industry? Perhaps the most important is the recognition that industry is not homogeneous but highly differentiated: the present research has identified four types, each with its own distinct profile. In order to implement an effective design policy the structure and culture of the type must be taken into account. Policies that are typologically inappropriate for a company will not succeed: global policies indiscriminately applied will create confusion and further reinforce the diffuse status of design within industry.

Another important implication follows from the recognition that there are clear attitudinal differences in those companies with a design manager. While these give rise to a more positive attitude to design, it is notable that there is no apparent difference in the organizational structure of these companies. This is rather like having a marketing function without the necessary organizational structure to implement it effectively. If the benefits generated by a design manager are to be realized, the organizational structure must be devised to profit from it: at present, this is not so. Also, when a company makes a commitment to design this represents a change in strategic focus. This must be managed as professionally as any other change and will almost certainly need management development and tolerance during the period of instability. A further implication derives from training: it is clear that designers – those trained in design schools as distinct from design engineers – hold beliefs about the role and extent of the design function that are significantly different from any other group. This difference, however, is less surprising than at first might appear. Designers tend to be treated differently within companies to marketing, production and finance managers. If designers were given the same range of company experience, training and responsibilities then their perceptions of the company, its procedures, functions and mission, could be expected to change. Designers are probably atypical because they are perceived as such and therefore treated as such – a self-fulfilling prophecy. If designers are to make an effective contribution to the management of a company then they should understand the various facets of the company and acquire the necessary functional breadth. If design is to be an integrated function within British industry then the designers must understand their industry.

The above can be encapsulated in three simple maxims:

1. A design policy cannot be purchased off-the-peg: it must be tailor-made.
2. Do not expect a design policy to be effective if the structure does not exist to implement it.
3. Do not expect designers to understand the company if the company does not develop methods to integrate them.

REFERENCES

1. NEDO, *Design for Corporate Culture*, National Economic Development Office, London, 1987.
2. Lorenz, C., 'Design in British Industry', *Financial Times*, 21 July 1987.
3. Gorb, P. and Dumas, A. M. P., 'Silent Design', *Design Studies*, July 1987.
4. *Ibid.*
5. Tichy, N. M., *Managing Strategic Change*, John Wiley, New York, 1983.
6. Zevulen, E., 'Multidimensional Scalogram Analysis: The Method and its Application', in Shye, S. (ed.), *Theory Construction and Data Analysis in the Behavioural Sciences*, Jossey-Bass, San Francisco, 1978.

*Ivor Owen is the director of
the Design Council. His
paper describes the problems
that exist in ensuring that
design is used effectively to
improve the industrial
performance of this country.
He then goes on to suggest
how the Design Council*

Industry and Design

plans to tackle these problems.

'Industry and Design' are two of the most important subjects we can debate in this country, both of which are inextricably connected. This paper is about some views of the industrial scene and how design is involved and also about how we in the Design Council see our future role, and how we hope to influence things. However, first let me explain my qualifications to talk to you about design.

I started by industrial career as a designer in the machine tool industry. I remember I never wanted to do anything else and I used to eat and sleep machine tool design. But by the time I was in my late twenties, I realized that it would be impossible for me to go through life being a designer. I also knew that if I was a German of American or Japanese – although I – none of us – knew much about Japan then in the 50s – I could have made a career as a designer if I so wished.

Two things made their impression on me then.

1. The combination of low status and low reward as an engineer or designer made it hopeless, long-term, as a career. We also had a number of German, Italian or Swiss working with us and it was clear

that at work they were regarded as an important indeed vital part of the organization, whereas we were regarded as a regrettably necessary overhead. People doffed their caps to engineers and DipEng was something to be admired and aspired to. Their standing in society was high and at work they were addressed as Herr DipEng.

2. It was obvious that our competitors in Europe clearly understood the importance of industrial design. Any attempt on our part to take such an approach was met by threats of instant dismissal on the grounds that such frippery was not necessary; it would increase costs when we were already unprofitable and that we should just concentrate on designs that worked. Indeed, we were arrogant about the technical performance to the point where we thought we had little to learn.

So I saw that the combination of the low status of engineers and the schism between engineering design and industrial design gave little hope for the machine tool industry. For the next thirty years I watched my views as a young man progressively become confirmed.

I moved fairly quickly into general management and in my early thirties became Manager of a large turbine factory making water and steam turbines, generators, nuclear reactor components and a wide variety of general engineering products. I then went into the computer industry, then into the bearing industry until I joined Thorn EMI in 1981 where I ran a large group of companies making a variety of products from machine tools to security systems and then later into consumer products and lighting.

And so for the middle part of my career, that is until 1981, I ran a succession of businesses which had a variety of things in common.

1. They were all at the leading edge of technology in one form or another, either in terms of design, manufacture or materials or processes (although that only probably applied to parts of the bearing business).

2. They were all unforgiving products. It is not possible to be a steam turbine manufacturer and perform badly. If turbine is not designed and made property to the right quality levels, it will not perform to specification and it might even explode. By the way, unlike the

appliance business, the automative business and many other businesses, it is not possible to perform badly and decline and die slowly
by losing market share.
3.   They were all very capital-intensive.
4.   It was very difficult to make a satisfactory profit.
5.   All one's competitors were uniformly extremely competent.

I used to long for the opportunity to differentiate ourselves against
our competitors by design, flair, quality, imagination, service to the
customer or some way of being different to our customers when usually
the only way was by some technological breakthrough or in terms of
materials or manufacturing technologies and capacity, all of which
usually needed very large capital sums, where there was little flexibility
and generally improvement came slowly – although trouble could come
very quickly. Indeed we did find, particularly in the bearing industry,
ways of being different, usually based on fitness of purpose based on
good engineering design and service which in fact convinced me that
there was no such thing as a commodity product even though at the time,
during the 70s, it was very much the perceived wisdom of the major
business schools that there was and that their simplistic but intellectually
arrogant solutions were the only solutions.

Later, when I became involved in the 80s in a range of products in
companies which were performing badly, I was stunned and horrified
at the lost opportunities which I could see because the products were
not well-designed, because design and its link with the other disciplines
involved with running a business, was not understood. Where it was also
true to say that the opportunity to reposition these businesses and, in
many cases, to stem import penetration, could be done relatively quickly
and relatively cheaply compared to some of the other businesses I had
been involved in and certainly without anything that could be described
as massive expenditure.

So what do we mean by 'design'? A little while ago, in searching for
the succinct definition, I said I thought it was a statement about customers, products and employees. It certainly is an all-pervasive subject.

Perhaps we could think of some of the aspects of design. All aspects
are interlinked and very few will work without the others. It starts, if we

are speaking of products, with a high standard of engineering design. That the fitness for purpose is correct; to a high quality and reliability levels and to the right cost levels. I also believe that the industrial design standards of all products must be high and this varies between different products. In appliances, say, it is a more obvious point than it is in others but, nevertheless, I can think of few products that are excluded from a need to think about industrial design.

This even includes steam turbines and motors and pumps and a huge range of engineering products. They all need an industrial design input. I strongly believe that the schism between engineering design and industrial design has been one of the most damaging issues in manufacturing industry imaginable.

Design also starts with an obsession with the customer and understanding what the customer wants. All too often this is not researched adequately and sometimes not at all. All too often there is almost no real understanding of what the customer thinks. There is no contact between the designer and the end user and there is precious little, very often, from a marketing or sales force who may deal with the distributor or the retailer but do not really understand how the customer thinks.

When I became chairman of a large appliance business, I went to work as a sales assistant in a high street retailer. I was convinced that the perceived wisdom of both the sales organization and the retailers that there were fixed price points beyond which it was impossible to sell a cooker, a kettle or an iron, was nonsense. I think retailers have been very damaging to the cause of good design and it is one of the reasons why you cannot get a variety of consumer products from British industry if you want top-level quality and specification and performance. I think all designers should spend some time in the customer's premises working direct with the customer, either working in a shop or working in a hotel kitchen if they are making commercial catering equipment. How often does an engine designer spend a few days in a service bay?

The next aspect of design is the environment in which it is made. I have always been adamant that well-designed, innovative products of good quality can only be made in factories and offices which also reach high standards. You cannot make such products in a slum. For the most part, British factories are scruffy compared to many of their international counterparts. We cannot, in fact, exclude the general environment. Generally our public environment is squalid and it has been steadily growing worse. Our roadside verges are not cut, and few people seem to be litter-conscious. However, I believe these things are connected and any visitor to Japan will see the point and one can soon see a hugely different approach to general standards in the USA and large parts of the Continent. To design and then make good products, the people who make them have to be well-treated, and what goes on outside matters, because people do not leave bad habits at the gate.

Then of course, design is about the management of design which again includes many broad issues. It starts with my earlier point about the reward and status of designers. Recently I have seen a designer in a multinational company who told me that when he was transferred to Germany, his salary increased three-and-a-half times. Status and reward to designers and engineers in industry in this country is still generally as far behind our main competitors now as it was forty years ago. It includes how product specifications are decided, and whether product development and product issues are ever discussed at board level. At what level is design in any way represented at board level and indeed should it be? I happen to think that the argument that there should be a director of design on the board of companies is not right. How do the various parts of the organization fit together? Simple questions like what does the customer want, what does the marketing director want? Are they understood at the time of debating the specification and the prototype development? You would hardly think that the answer was 'no' but in very many cases that is in fact the case.

If design is anything, then it is of course multi-disciplinary starting with marketing. In many cases it is a toss-up whether marketing comes first or good product design comes first. Either way, there must be no schism between the two any more than there should be between industrial design and engineering design. It includes manufacturing technology, innovation, material technology.

And then there are what I call downstream issues starting with the generally inhospitable ethos in this country and the frequent anti-industry bias in education and institutions. There are huge problems in our educational system. There are now real signs of change but there is still a long, long way to go. Industry presents itself badly – young people do not find industry attractive. There is a massive shortage of engineers with the correct training. Yet places are being reduced and only 50 per cent of those who read engineering go into it.

Our inadequate skill base is a matter for great concern. We seem to be very bad at avoiding making people feel failures if they do not get a certain number of 'O' and 'A' levels and then go on to university. It seems beyond us to generate a general attitude of respect and aspiration for vocational skills.

In other countries that is different and on the Continent they are pouring out people with vocational skills in massively greater numbers than we are, and indeed, in a number of industries, it is now obvious that where we can design and develop the products, we cannot make them because we simply do not have the skill base. I remember years ago when a civil servant complained to me about the performance of British industry and I remember saying 'What do you expect when the whole institutional system encourages the best brains to keep away from industry and indeed educates and trains them in ways that are unsuitable for industry?'

Why do the small European countries have so many major international players? Holland and Sweden are small countries compared to this country but they have Philips, Electrolux, Volvo, SKF, Saab. Frequently, messages come from senior government ministers that they do not think that a strong manufacturing base in this country is important. It was recently said that increasing imports were good news because it showed that industry was re-equipping with plant and equipment but are we the only country that does not understand that if, in the long term, you want to have a manufacturing base, then you have to make the equipment needed to manufacture your products? I remember, certainly in the early 60s, if we ever wanted to buy a piece of equipment that was not of British manufacture, there was a terrific song and dance and approval was only ever given if there was clearly no alternative. The decline of our industrial base and the ever-increasing tide of imports, in my experience, put a stop to that and now I always find it depressing if you look around any renovated or new factory to see that a high proportion of equipment in it from machine tools to air-conditioning plant to paint plant or whatever, is from a foreign manufacturer; the same goes increasingly for buildings.

You may think these issues are a long way from any debate about design. There are instances where a narrow debate about design is in itself all that is needed and, even allowing that it has to be reasonably broad to cover some of the points I have made before, like industrial design, marketing or whatever. But for the most part, in one degree or other, all the issues I have touched upon become relevant to a debate about the success or not of businesses or industries through good design. Indeed, the vital importance of design and the ability of it to contribute to the profitable performance of companies, has often lost credibility because of the simplistic message – take a good dose of design and all will be well. It is true to say that no business will be successful unless its products are well-designed, but it is not true to say that well-designed products will make a good business. Even with a modicum of understanding of the word, no one would seriously pretend that it would.

I would now like to turn to the general economic scene for a moment and then to the Design Council and its future role in all this. I believe that it is nonsense to ever believe that an expanding service industry can replace a declining manufacturing industry. In 1955 manufacturing exports were two-and-a-half times as valuable as manufacturing imports. In 1982 they were in balance and now our manufacturing trade deficit is approaching £20 billion. When the income from our overseas assets is excluded, our surplus on invisibles is very small and the potential scope for improvement in this area is not in the same order of magnitude as the manufacturing deficit. As for balance of payments deficits, they have just moved into an overall deficit on invisibles.

Many weaknesses other than design have contributed to the long-term, relative decline of UK manufacturing. These include lack of enterprise, weak marketing, investment deficiencies, shortage of technical skills, poor industrial relations and the generally unsupportive national ethos which looks down upon and undervalues the manufacturing industry. Many of these factors have already been significantly addressed in recent years but uncompetitive product design remains an important contributory source of the national deficiency and will require substantial further effort over a long timescale to resolve.

The present trade deficit is more serious than often realized for the following reasons.

1. Compared with our more successful industrial competitors among advanced countries, much of our manufacturing output is of the semi-commodity, price-sensitive variety open to increasing competition from the developing world.

2. The UK's manufacturing trade problems now lie mainly in non-price factors such as product quality, design and the loss of key skills and infrastructures. It is clearly shown by the fact that most of our imports are from higher-cost countries, with West Germany being particularly prominent. If you look around at high added value products with high design content, then increasingly we do not compete. The next time you go for an eye check-up or to the dentist, have a look at the very sophisticated equipment which he or she has and the majority of which comes from overseas. My local optician's equipment is German or Japanese with one exception, which is a piece of British equipment which recently won a Design Council Award, the Pulsair Tonometer. It may be a digression but I know that whenever I look at advanced equipment from overseas, I think of other factors which will influence performance. Profit among Japanese companies is much lower compared to those in the UK but it is easier for them to do so with very low interest charges and ready availability of capital. Equally, German companies tend to be financed by banks rather than equity and an approach to the handling of their shareholders is very different from ours. Despite problems in North America, where the stock market is ruthless on short-term performance, they managed to operate long-term policies of research and development in a way that we find difficult. All these factors may have to change before a real substantial design-led renaissance is possible.

In the increasingly competitive 90s, the average consumer will probably behave like the average West German consumer in the 70s and largely reject cheaper products. There is no shortage of evidence to show this happening substantially in consumer products and indeed we now must recognize that many consumer products have left these shores, or at least the ownership of British companies, forever and many of the growth of high-tech, high added value, high-income, high-investment products like video, fax, have never been here and are not likely to be.

I hope my arguments sound realistic and factual rather than pessimistic. The balance between the two is never easy and too much of one or the other can be very damaging. There has been a rekindling of business confidence and there has been a restructuring of Britain's industrial base in recent years and this has contributed much to the novel sense of economic well-being this country has been experiencing of late. In many areas, these changes in business and industry have translated into great improvement in competitive performance. A number of our industrial and manufacturing companies are now stars shining very brightly indeed in the markets of the world. There is a very different atmosphere generally towards design and productivity. Look at the strides made by the steel industry.

There is a strong debate in the education field about the importance of design and the Government's recognition of the need for a more positive attitude in education towards the requirements and status of industry and the prospect for a national curriculum are welcome as major opportunities for improving design education. We are also, for example, promoting the recognition of 'A'-level design courses for university entry qualification with increasing success. However, the overall improvement has not gone anything like as far or wide enough to give cause for national self-congratulation. Too many sectors and companies are still not competing sufficiently either to combat imports in the home market or to make a quality-led expansion of their exports. This is one of the factors that has contributed to the country's growing balance of payments deficit. But there is now a lot to build on and many basic problems of the past have been or are being addressed.

THE DESIGN COUNCIL

Therefore, do not envy the Design Council. This unique organization is uniquely placed to play a key role in improving the performance of British industry. There have been changes in the Design Council during this year and I would like to tell you about some of them and about our plans to play a constructive role in influencing the solution to all the issues I have touched upon.

Our activities are focused on achieving two key objectives.

1. To help British manufacturing industry develop better products through the total design process and to improve their competitive position in all markets.

2. To 'take action' to support and improve the quality, content and quantity of design education and training.

From this a number of things follow.

1. First, there is the need to focus and rationalize our activities and this has led to the cessation of a number of activities.

2. We stopped the Design Centre selection and label schemes earlier in the year, together with Marketing Services.

3. We have now announced the closure of the shop and café in the Haymarket, which will become a centre devoted to showing the work of young designers.

4. Similar programmes have been carried out in Scotland and Wales.

5. We are 30 per cent smaller than at the beginning of the year.

6. We are clear about our role which is to be a catalyst.

7. We recognize that we need to be as small as possible – not as large as possible – efficient and responsive.

8. From this it follows that we have to lever our messages by a variety of means but certainly we will be strong networkers, co-operating and working with and through other organizations.

9. Indeed, by definition, if we believe that design swiftly embraces many other disciplines, then it must be nonsense to believe that design 'belongs' to the Design Council alone and we must help other organizations to see that they cannot teach marketing, quality, manufacturing technology or even physics or history without design.

10. What we do will be measurable, in terms of perceived need, effectiveness or numerate criteria.

11. We recognize that we must not solely rely on our Grant-in-Aid to
    carry out our work and we will reduce our dependency on Govern-
    ment funding while expanding in certain directions – for example
    in training.

We will develop our regional presence using our Wolverhampton office
as a pilot for a role designed for the needs of the region and covering
education as well as services to industry. Exhibitions are an expensive
way of communicating succinct messages to a specified audience. We
shall become more campaigning. Where it is intelligent to do so, we shall
be controversial and, for example, show foreign products, something
that we have never done. We shall of course continue to work closely
with the DTI both as a contractor under the Enterprise Initiative and also
in developing campaigns, covering various aspects of design, for
example, the management of design. Which brings me back to the main
thrust of our work in industry and education.

We are now establishing a small research department which will
identify products or sectors where change through design is necessary
and possible in a significant way. But with an intelligent understanding
of the broad strategic issues and with an international view. We will
develop ways of providing services which probably may be charged for
but will range from the management of design through innovation to the
selection of materials and relate these programmes to the areas we focus
on through our research and this will be strongly linked with our regional
programme.

In education our principal aims are to ensure:
1. That design is a component of the education of all children between
   the ages of five and sixteen and that children have the opportunity
   to learn design within a variety of contexts within the curriculum.
2. That higher education institutions recognize success in design 'A'
   levels as a valid qualification for entry to courses and make it a
   prerequisite for entry to design-based courses.
3. That degree and diploma courses in industrial design reflect the
   requirements of industry, and that there is adequate communication
   between colleges and designers and industrialists.
4. That both undergraduate courses and research in engineering con-

tain a significant and appropriate design element and that an appropriate academic and industrial framework exists for the development of engineering design education.

Underpinning this work we have some marvellous publications. *The Big Paper* is a regular paper for primary school teachers. It describes successful design work going on in primary schools and also carries a range of features material, some of which can be used directly with children. *Designing* is aimed at secondary schools and can be used equally by teachers and students. It covers the whole range of design activity – from stained glass to engineering – and, like *The Big Paper*, communicates with schools in a stimulating and visually attractive way. Our third regular publication, *Engineering Design Education and Training*, attempts to bring home some of the realities of design to those studying and teaching engineering at universities and colleges, and is now increasingly aimed at those responsible for experience in industry. An exciting new project we are working on is a paper for design students in colleges and polytechnics, showing exciting and innovative work going on in these places, especially work relevant to industrial needs.

In addition to these serial publications, we have produced a number of videos for use as teaching and learning resources in schools and colleges. The quality of these is shown by the fact that the last two have won prizes at the New York Film Festival. In fact there is no limit to the range of educational material we could publish and we are continually looking for new opportunities.

CONCLUSION

Turning round the performance of British industry to the point where it again makes a positive contribution to the nation's balance of payments is a huge task. It will take the concerted efforts of all government departments and agencies involved, acting coherently and to a donsistent, realistic and clearly defined policy, to make any deep and lasting impact on the problem.

Companies with price and commodity-sensitive exports must be persuaded to shift their emphasis to quality and special market appeal in order to free themselves from the effects of the strong pound. We know it can be done. Other countries with strong currencies run manufacturing surpluses, maintaining them even as their currencies rise. Their goods have qualities for which there is a demand even when they are not the cheapest. Britain does have potential for a thorough manufacturing revival to build on the improvements which have taken place over the last few years where goods compete primarily on special quality and design. Much hard work has already been done to this end but it has only got us a little way off the starting line.

The Design Council has an important part to play in this overall effort which also involves investment, marketing, productivity, industrial relations and overall changes in attitudes to enterprise and the status of industry.

I have tried in this paper to present a brief picture of some of the issues as I see them, which need to form a background to any debate about industry and design. I hope that I have also shown you what an exciting position the Design Council finds itself in, very excited at the prospect of playing a key role in the challenge and the dramatic improvements I see ahead.

*Lord Gowrie is the Chairman
of Sotheby's, the leading
international auctioneers of
fine art. Lord Gowrie was a
former Minister for the Arts
and Treasury spokesman for
the House of Lords. He is
Provost of the Royal College
of Art and has published
widely in the field of the fine
arts and English literature.*

The Social Psychology of Design

I want to start by saying that I am more than conventionally honoured to have been asked to talk to you for a few moments on the subject of design. My title might be 'some thoughts on the social psychology of design'. I do not mean that too literally, however, because this evening is a seminar. Your contribution later will be more important than my own.

If I can get the ball rolling I shall be more than happy. I am also honoured because the London Business School, and Peter Gorb's design unit in particular, has taken the lead in recent years in bringing home to British industry the message that good design is good business. It is a commercial and economic imperative. The developed world and the open societies are at peace and they are also operating still under a liberal trade regimen, that is to say without too-great restrictions on free trade. I do not imagine that President Bush is listening to me but if he is, long may it continue. In such a world both the demand for goods and the nature of the goods which find themselves in demand are in fact design-led. Design conditions to a large degree what is sold rather than the other way round.

Take two obvious examples – cars and clothes. Both have, to use the language appropriate to each, long overtaken or outstripped the need for transportation or the dictates of warmth or modesty. These are products which human beings make and sell with an idea, some would call it a fantasy, in mind. A great 20th-century poet, Yeats, wrote a line which has always haunted me: 'In dreams begin responsibilities'.

Now there is a theme for us. I believe that design (and here I look rather nervously towards Stephen Bayley and behind him Terence Conran) has relatively little to do with utility. Corbusier's dictum that form is function is out of date, if ever it was in date. I suggest that designers need to pay a great deal of attention to ideas, to dreams, to fantasies, to social psychology, to history – taking history as being the way our ideas of time past shape our behaviour in time present. History does not really exist – it is what a given time thinks may have happened in a previous time.

I want to suggest that designers cannot really rest on an idea of themselves as down-to-earth, sensible people, who know a good thing when they see one. They need to understand the importance of poetic or metaphorical truth where human beings are concerned. I am some-what out of touch as a teacher – I used to teach at University College London fifteen years ago – but I understand that the study of the collective unconscious has been out of fashion recently. It seems to me that advertising men are at least as conditioned by a reading of Jung as by a reading of Freud, though if part of the point of good design is to make people want things, you need to understand, in a rather Freudian way, the forces that condition wanting, the dreams that compel our desires. This is of course very dangerous knowledge and it is highly corruptible – we have to face that. I shall try to suggest some ways of dealing with the ethics of the thing as I go along, and you will have ideas about it too.

But how can we invent a way of designing responsibilities, so to say, into any job specification before us without ruining the fun or the excitement or the buzz of the thing? That is the challenge that faces contemporary designers. I am well aware of the enormity of what I am suggesting so let me make a few quick and again rather conventional disclaimers.

I am neither an industrialist, nor an artist, nor a designer. You have heard of Yuppies, and I am too old to be one of them. I think of myself as a PIC which I take to be, in my own slang, a Partially Informed Consumer. That is to say I am the kind of person whom people pitch design at because I am the kind of person who cares about it, who minds very much how things look. And the social psychology of people like myself is really what a lot of designers have in mind, in my view. It happens that I myself work these days in a service industry which deals in works of art and in the decorative objects of the past. Most of the things that I sell are useless and few have real value – that is to say, in strict economic terms they do not generate income. You cannot really own a painting it seems to me: you can only take a life interest or leasehold over it. I do know, therefore, from this employment, as well as from the dictates of my own temperament, that sensuous response to the visible and the tactile is a great force in human life. It pushes and pulls us in different and often uncomfortable directions. Our ideas, for instance, of understanding, of perception, are intimately connected with the language of the eye, with the language of seeing – the words we use show that. And what is our imagination without images?

Of course the great propeller of sex has a lot to do with design. In what are euphemstically called 'men's magazines' the bosoms of sleek untouchable girls share space with bonnets and bumpers of sleek unaffordable automobiles. I am a rotten driver, I hate speed and I have no mechanical understanding, yet I think that if I were a rich man I would buy a Ferrari because it seems to me the most beautiful object I have seen in recent years in terms of exterior design at least. It is pretty nearly useless, of course, in that it is uncomfortable as transportation for more than one or two people; it is not designed as a racing car and yet it can travel at racing speeds which are either illegal or impractical. What it is designed to do is trade on dreams of power over nature and power over mortality. Now dreams of this kind, it seems to me, are not restricted to males but they are intimately related to male sexuality. My belief is that there are moral, political and practical forces at work in the contemporary world at present which are making the Testeross, fine object though it is, a bit out of date, which are impelling us as a species towards dreams more intimately connected with the female psyche. If I am right, this will have profound effects on design. Crudely put, female sexuality is less concerned with power and domination than male. It is more conservative and more conserving. Mankind has almost universally treated the earth as being female and the sky as male. God, you will remember, is a bloke and lives in the sky. An American poet of my acquaintance put this point rather pithily:

Higgledy, piggledy,
Jupiter Pluvius,
ante-diluvius
rain-making chap,

showered his masculine
potentialities
all over Danäe's
succulent lap.

The Prince of Wales has been much in the news recently as a man interested in the way things look and as an opinion-former concerned that we should all get more interested in the way things look. He seems to be a man sufficiently confident of his own masculinity (remember all those Action Man jokes and the windsurfing and the skiing and the polo playing, and the rest of it) to use his excellent mind and communication skills as well as his public position to draw our attention to the need to conserve the earth and its resources, to the habitat in the local, micro-economic sense of the word: home, hearth, village and community. Modern technology and planning techniques in fact make this a relatively easy thing to do. It is not difficult to design houses that look like a child's idea of a house with the hat for a roof and eyes for windows, rather than an architect's idea of architecture or an accountant's idea of profit. However, the power principle, or masculine principle, is less strong in such an idea of a house than in tower blocks. To quote another line of doggerel: 'God forgive the phallic flashers who built the Barbican.'

When we talk about human scale we are employing the female side of our sensibility. It may be that the emergent power in the latter part of the 20th century of women as workers, consumers and voters, is behind our increased concern for nature. Here is an area where design and ethics seem to me to meet. Our ideas of beauty meet our ideas of truth.

I want to go on stressing this use of the word idea because I am still, of course, trafficking in ideas rather than products. There is a strong philosophical case, and I must acknowledge this, that nature herself, itself, is not masculine or feminine but neutral. People think of the country as 'nature' contrasting it with the town. But as you all know, what we call 'the country' is a human artefact, an aggregate of designs, of uses. You do not need to explain this to gardeners. 'Purposeless matter hovering in the dark', another line of poetry, is what we run into with nature when we forget to weed.

Mankind may be threatened as a species by nuclear weapons, but I myself am very sceptical as to whether nature is. I understand that contemporary geologists believe that evolution was interrupted for about 60 million years by some ecocatastrophe like a very large meteorite hitting the earth. It was interrupted and modified but it was not stopped in its tracks.

In case you may at this point be thinking that I am getting like my hobby – too airy-fairy or philosophical for a business-school talk, let me suggest how some of these archetypical design motives might translate into design motifs.

Let me get down to the brass tacks of money. I think everyone here, looking cursorarily round, will have had first hand experience of the liberating decade of the 60s when there was a new democracy of design. In Britain in particular, and in Swinging London in Britain, bounds and gestures were pushed to one side (in Philip Larkin's phrase) like an outdated combine harvester. But before that, the arbiters of taste were, here as elsewhere, mainly the upper bourgeoisie and those who served them commercially.

Growing up in the 50s, mostly in a military town, I remember that the glass of fashion and the mould of form was the young Cavalry or Brigade of Guards Officer. His uniforms; his town wear with bowler hat, rolled umbrella and suit; his flannels, hunting jacket, polo clothes; his shooting brake; his guns; his Alvis or his Bentley – all were part of his magic. The merchants who served him would even make up special cigarettes. You will remember James Bond – Ian Fleming's James Bond, not the Broccolli brothers' James Bond – who was not out of place in White's Club and who smoked his own brand of cigarettes. In my town, for instance, the shoe merchants would place leather brogues in the sunny windows of their shops in order that they would age unevenly, exactly the right taste specification for this market.

To most of us in this room there is a piece of British nostalgia, quaint in the context of a class system whose diminishing influence most of us welcome. However, I must tell you that in Yves St Laurent, who first adapted this style, for day clothes for women, or to Ralph Lauren, who exploits it today – it is nothing to do with class. It is a gigantic business. Lauren shops are exact cribs of the old upper-class merchant interiors of my youth and the gun or billiard rooms of the stately homes. They, all over the world, are making millions out of one part of our style. We decided to go in for something cooler and more Scandinavian in influence. Useful, practical, classless and, dare I say it rather dull – much less concerned with fantasy or with dreams. So please never underestimate the role of nostalgia in design or in marketing. It was not just Marie Antoinette who dressed to the nines as a peasant, it is everyone today who buys a 'farmhouse' kitchen.

When did you last smoke a ham in your cottage or flat near the Fulham Road? However, you can in fact buy the clobber to do so in shops up and down it. The French Impressionists thought that the most beautiful building in London (they came here to dodge the Franco-Prussian War) was the House of Parliament. I happen to agree with them. Those infinitely crenellated surfaces take the smoky and misty Thames-side light better than any other building. Barry was faking high Gothic but it seems to me that he produced something pretty well as good as his medieval models, in this country at least. British neo-Classicism, the preferred style a hundred years before Barry and still a great determining theme in establishment style in Britain today, is in fact neo-Palladian; Palladio is neo-Roman; Roman is neo-Greek. Revolutionary changes in style and sensibility do take place, but they are extremely rare. I would guess they probably only occur two or three times in a millennium.

The Cubism-Constructivism of the early part of this century is one such revolution and, speaking for myself, when I see a contemporary building which I happen to think good, it is usually neo-Art Deco, Art Deco being the style that grew out of Cubism and Constructivism. A fine example is the Chapman Taylor Building behind Vauxhall Bridge.

As I suggested earlier, I believe it may be possible, if peace prevails, that a more feminine sensibility may be about to determine Western culture as we approach the new millennium. I include within this the feminine side of masculine sensibilities because, thank God, our minds and bodies harbour both genders. In a way which I have not begun to work out, but do nevertheless intuit, I find the growth of information technologies supportive of this view. There is something compact, sealed, functional yet secret, something feminine about computers. Hobbies are also a giveaway. I am fascinated by how many men nowadays read decorative magazines. At Sotheby's marketing department, with the male writer of cheques still uppermost in our minds, we have taken note of this.

Even the blue Tories are getting a lot of mileage out of green politics. Mr Kinnock, who has reason to be a bit aggrieved about this, is liable to founder here as there is no great apotheosis, in my experience, of collective masculinity than the TUC. Men are dressing down nowadays – women are dressing up. There is pressure on men to look after their health and figures. There is pressure on mankind to look after its world. Economic growth is coming from the exploitation of ideas and dreams rather than the exploitation of raw materials.

All these factors are beginning to affect marketing and design. I talked perhaps more than I should have done about poetry and economics but please remember that poetry derives from the Greek word to make and economics is the Greek word for household. Our household, our habitat, is what we make of it. The ethical imperative is that we only make well what we love and when we love, and what we have done

and what we see we have done teaches us how to love. Let me close
with one more bit of doggerel, but this time by a great poet:

Love requires an object
And this varies so much
Almost, I imagine,
Anything will do.
When I was a child I
Loved a pumping engine:
Thought it every bit as
Beautiful as you.

Stephen Bayley was until recently the Chief Executive of the Design Museum in Docklands with which he has been associated since its inception as the Boilerhouse at the Victoria and Albert Museum. Prior to that he was a member of the academic staff at the University of Kent where he taught design history, a subject in which he has written extensively.

Design, Commerce and Culture

When I was a child I used to read architecture magazines because I was fascinated by modern interiors. At school in a magnificently depressing northern sandstone Gothic pile with dark wood and stained glass windows, *Architectural Review* or *House and Garden* presented me with visions of clean, rational, bright alternatives which I had hitherto only glimpsed on the 405 lines of 'Seventy-Seven Sunset Strip'. The fascination was more than mere curiosity, since I craved those environments as repenters crave salvation. They provoked an intense longing. Certainly, there was a lot of emotional energy swirling around my adolescent reading.

So, when I was flicking through features illustrating Helsinki apartments or exhibitions in Milan or a new airport in Texas, the atmosphere was highly charged. I was receptive, my imagination a suction pump. It was in the same journals that I first came across the people responsible for these interiors of desire, the designers. I do not mean the well-intentioned, homespun, Cotswold artisan craftsmen, but colourful exotics with far more appeal to a boy with wandering eye in mind.

COMMERCE CULTURE AND THE FUTURE

The Design Museum where I work has looked at a wide range of artistic, social, technical and cultural ideas which have formed the climate of thought in which designers work. But what of the future? Now that the post-modern deregulation of history has demolished the old conviction that for any age there was a single, unifying style which expressed the spirit of the age, now that 'designer' is as pejorative a term as 'executive' or 'de luxe', now that all available aesthetic choice is to be found on somebody or other's computerized expert system, what scope is there for designers?

There used to be features showing sleek, pomaded, mustachioed creatives like Raymond Loewy. I first saw him circa 1962, dressed in a white suit, extravagant foulard and spats, posing by the new Studebaker Avanti whose striking body he had just designed. Or Charles Eames, handsome, square-jawed, tough, but sensitive, drop-dead cool. He was in a wash'n'wear plaid shirt, sitting in the artistically cluttered studio of his Venice, California, house. This habitation he made from industrial components. Then take Eliot Noyes, the very picture of New England refinement with his crew cut, his button-downs and his Beechcraft. Noyes was the Harvard architect who redesigned the entire appearance of IBM, telling the chairman in a memorably pithy memorandum announcing his proposals 'You would prefer neatness'. In the American magazines, Noyes was frequently photographed with his wife, Molly, near their beautiful Connecticut house. They always used to have matching cars: this year Thunderbirds, then Porsche 356s, even Land-Rovers.

And all of this was marvellous, giving a clear impression that design was all about shaping appearance, about changing the look of things. And so in a way it still is, but it has become other things as well; and because of that the 'look of things' has moved somewhere else in the system of values where we judge the quality and character of those things.

When people talk about design they are talking – more or less consciously – about two things. The first is the simplest to grasp. Design is what a certain group of professionals and artisans do when they engage in making creative decisions about the function and appearance of the things we buy and use. The second is more abstract, but places design on a par with, say, literature and the fine arts in its status as an activity which defines man's relationship with the material world.

This is design as an activity which embraces both production and consumption, as it were, the critical intelligence co-ordinating supply and demand. This activity involves both professional designers, manufacturers and – perhaps most importantly – consumers. Few histories involving design have been written, which is not to say that histories of what designers have done are lacking. Their work at the Museum is an attempt to lay the basis for such a history so that in future a history of civilization might be written which is as much concerned with objects as well as with wars.

ART AND DESIGN

The Design Museum is also concerned with the relationship of art with the everyday in order to understand what influences consumer choice in a world of rapidly changing technical and social values. This is why much of its early work was concerned with examining a work of art created as an artefact, and went on to question what real merit lies in artistic traditions when modern technology undermines their cultural basis. Similarly, what deeper structure is suggested by our entrenched attitudes to reproductions?

Museums and department stores are among the most fascinating institutional creations of the last century and a half. By putting the merchandise of the world, past and present, on display they fomented a crisis which has led to the banalization of art. This crisis culminated in a merger of the two. One other manifestation of the same crisis was the designer cult, an expression of the infantile stage of the history of industrial design.

DESIGNING TIME AND SPACE

Nowadays, design is not simply about appearance, but is increasingly concerned with . . . experience; and the most important experience of the 20th century is speed. Indeed, as Aldous Huxley once remarked, speed is in fact the only entirely novel sensation of our age. Flight, after all, was known in the 18th century, if only to the Montgolfier brothers and whoever shared their pillow talk. And speed is exciting: in his 1905 essay on infant sexuality Freud explained that children, especially boys, get particular excitation from sensations of movement, hence the historically specific engine driver fantasy.

Speed compresses time and space. You can have a lot of fun compressing time and space in a Ferrari, but, quite frankly, it is from the infantile stage of industrial production and does not have all that much to do with the future development of Western material culture. Speed in a more metaphorical context most certainly does.

This thought came to me a few months ago when I was hanging around the edges of a conversation between some big names in the motor industry. Eaves-dropping ten years ago would have been all about design: the industry men would have been muttering knowingly about its importance, its contribution to the bottom line, to perceived value and all of that. Nowadays they know it is important and confer on it that greatest of compliments, of being taken for granted; they would be no more likely to have a heated exchange of views about design than they would about any other accepted fact of industrialized capitalist production, such as discounted cash flow or inventory control.

DISTRIBUTION

Instead, the know-all mutterings today are about distribution. Now that functional and aesthetic excellence are the baseline for any company that expects to survive beyond the end of the week, the real competitive advantage comes from mastery of time and space. You speak of Fiat or Ford or Toyota or Toshiba or Benetton and you will find this is what concerns them most.

In its Castrette warehouse Benetton owns the most sophisticated distribution centre in the world, serving the world from one building. Benetton does not manufacture a single garment until it is ordered and just as soon as the order is processed your woolly is documented and tracked through the entire manufacturing process all the way into the 280,000-cubic-metre warehouse where Comau robots store it and its sibling woollies in a random access system not dissimilar in its complexity and efficiency to a computer's memory.

To the human visitor the system looks bewildering, with boxes for Chelmsford hugger-mugger with boxes for Tokyo, but the computer knows where everything is, sees it onto conveyors, into trucks and speedily to market. Benetton says the whole process takes no more than seventeen days, of which a rigid seven are accounted for by distribution irrespective of destination. This system keeps Benetton in constant, direct contact with its stores and turns on their head all the old assumptions about design-for-industry because now the possibility exists, within that amazingly short seventeen-day cycle, for designer and manufacturer to be directly responsive to changing public taste. Soon they might start responding, rather than imposing.

TELESHOPPING

Since museums and department stores first put the man-made world on display, the public has demanded of all new technologies that they provide a similar service. First it was catalogue showrooms on the Wedgwood model, then it was department stores, then it was the telephone which changed experience.[1] In *Automation* (1952) John Diebold[2] explained that all new technologies, from print through to artificial intelligence, tend to impose the following pattern of changes: *(a)* allow you to mechanize what you did yesterday; *(b)* but then you find technology changes the actual task; *(c)* and then changes in society follow.

The latest innovation is teleshopping, a phenomenon that brings the discussion of commerce and culture full circle. Teleshopping has the potential to realize in full the cultural opportunities suggested in André

Malraux's idea of a future 'museum without walls' while satisfying the commercial appetites of Aristide Boucicaut's Bon Marché. Presently restricted to test markets in the USA, teleshopping could bring the museum and the store into your own home.

Teleshopping depends not only on the consumer's appetites, but also on two new technologies: fibrox[3] and the fast-approaching integration of computers and video. Presently, telecommunication companies are laying fibrox cable for voice and data transmission, but they are anxious to acquire 'entertainment' too, since the capacity of fibrox allows the carriage of video as well as voice and numbers.

At the same time television receivers are slowly becoming integrated with computers. At the moment they have something in common, both televisions and most computer screens using old-fashioned cathode ray tubes, but by the early 90s the new machines will use new technology to combine high-definition image processing with vast computing power.[4] The TV computers will work through fibrox, carrying not only 'traditional television' but also educational and commercial material.[5] However, other applications of speed will have a crucial influence on design.

GOD IS IN THE FAST LANE

For about three decades after 1945 the Japanese devoted most of their considerable ingenuity and organizational energy to process efficiency rather than innovation. This characteristically long-term view has now endowed them with an investment of factories so bewilderingly efficient that creativity actually comes out of shop-floor. It is like a calculus: the Japanese have some brilliant designs not simply because they have brilliant designers, but because they can manufacture anything imaginable.

In Europe even the most advanced manufacturing companies, Volkswagen for instance, have product life cycles of about eight years. The Japanese motor industry is approaching twenty-four months and electronic products are, in some cases, down to a matter of weeks. When you learn that Hitachi manufactures video recorders in ninety seconds, you realize that the Japanese are driving themselves up a helix of efficiency which actually forces creativity. You also realize that the examples of Raymond Loewy, Charles Eames and Eliot Noyes are as remote from contemporary issues as Grinling Gibbons or Hepplewhite.

Some Western manufacturers are catching up: Motorola now makes its pagers in two hours when before it used to take three weeks. Hewlett-Packard too has learnt the lesson of the Japanese. Chief Executive, John Young, says: 'Doing it fast forces you to do it right first time.'

Thus industrial designers are exposed to Darwinian disciplines hitherto more familiar to racing drivers. With this new emphasis on distribution, time, space and the innovation cycle, aesthetics seem as quaint as *ars longa, vita brevis* did to General Motors in the 50s . . . or is this going to far, too fast?

The old focused transfer lines and the principles of linear production established by Henry Ford produced a very particular aesthetic; designers reacted to the disciplines of mass production by developing a visual language whose vocabulary included things like cut lines, proportion, raddii, mouldings. However, the new industrial divinity is not manufacturing, but speed. Eventually, when the awareness of this reshuffle seeps into culture, the divinity will have idols made in its image. God used to be in the details, but now he is in the fast lane. When he pulls in for a breather, the Loewy, Eames and Noyes of the next generation are going to be waiting for him.

CROSSOVERS

Certainly, industral design is going to change. Mario Bellini is one of Italy's greatest product designers. Astonishingly, he has said that the term 'industrial design' is redundant,[6] believing that it is futile to separate it from any other type of design activity, whether that is architecture, laying the table or hairdressing. The potency of technology allows designers to be purely creative, using the vast repository of material generated by thousands of years of civilization, and also potentially making new discoveries on the way. Bellini says that the product itself is now more important than the process.

'I think it is very positive this new crossing between pure art and applied art and design, because this interchange will nourish both the user's culture and the designer's culture.'

As an architect himself, Bellini is well aware that the most satisfactory 20th-century furniture was designed by architects. As a consultant to Olivetti he is equally aware of the momentous changes technology will impose on the organization of work . . . but not necessarily on life. On the contrary, the future seems to hold more and better opportunities for designers and Bellini is only speaking for the Spirit of the Age when he says what he most keenly anticipates is a general return to the 'culture of living in a house'.

REFERENCES

1.   In a curious article, 'Long Distance Shopping by Telephone', *The Daily Mail* for 28 April 1907 explained 'Many villages . . . are now equipped with a service, thus enabling women to order their household necessities at the great London stores and secure a country life with city conveniences.'

2.   John Diebold, by his own account, coined the word 'automation' when he actually failed to spell 'automatization' correctly.

3.   Fibro is an elision of 'fibre-optical', multi-cored, cable-carrying digitalized laser information. The present capacity of fibrox is about 1,000 times greater than copper wire.

4.   Nicholas Negroponte, of the Massachusetts Institute of Technology's Media Lab, says the typical home computer of the future will have the capacity of current Cray supercomputers, capable, for instance, of processing 50 million instructions per second (*Forbes*, 20 February 1989).

5.   The pharmaceutical company, Smith Kline Beckman, already uses fibrox for 'narrowcasting' of training films to doctors, pharmacists and salesmen.

6.   In *Design*, No. 481.

This is an edited version of an article originally commissioned for Design Management: A Handbook of Issues and Methods, *edited by Mark Oakley (Basil Blackwell, 1990).*

Design as a Corporate Weapon

In the continuously shifting focus of management preoccupations there is little doubt that in the UK at least, the 80s have become the decade of design.

These shifts are relatively slow, with a particular area of specialist attention often dominating for up to two decades, and the leading edge of change usually coming from the USA. In the 40s and early 50s production and productivity occupied central stage as the world finished its major wars and re-equipped with capital goods in both developed and developing countries. Ten of more years of focus on the behavioural sciences gave way in the early 60s to marketing, with the increasing need to stimulate worldwide consumer demand for the goods and services previously available to the wealthy in the developed world.

Not until the mid-70s did the Americans begin to worry about the competition from Japan which had begun to eat into their home markets. The focus was shifting again – this time to design and the leadership for the first time was coming outside the USA.

Apart from a few early pioneers the message about the importance of design to managers hit Britain with some force in the early 80s. The main agent of change was the support given by the Prime Minister who put both government money and manpower into promoting the design cause.[1] Industry and education were nudged into a design courtship, and by the middle of the decade the media, and television in particular, was busy celebrating the marriage. Design is now well-established as a powerful tool in the hands of managers who need to make and sell products or develop an effective working environment, or communicate efficiently to consumers, customers, shareholders and others. Most senior executives now expect to use design resources in the development and communication of business strategies. More and more MBA courses are adding design and its management to their curricula. The only remaining question is whether design as a key management preoccupation will continue to flourish through more than one decade, or indeed perhaps until the end of the century.

Fashion will of course be a major determinant. However, it needs more than fashion to keep an idea and an expertise on the commanding heights of management thinking. There is no doubt that permanent and responsible disciplines like marketing and operations management hold their place because of their abilities to embrace a range of long-lasting management preoccupations; and this will also need to be true of design and design management, if they are to last the course.

There have been many long-running management preoccupations, but four fairly permanent ones are as follows.

1. Innovation and place that the innovative process occupies in the search for profits.
2. Quality and its control and maintenance in both production and service organizations.
3. The prerequisites for developing effective line managers expressed in terms of education, training and development.
4. The design contribution to corporate strategy expressed through the field of corporate identity design.

These four will be a good starting point if we are to evaluate the permanent role that design and its management may have in the coming years. We will need to take a look at how design relates and contributes to each of them.

However, before we begin to do this we need to clear some ground. We need to be sure of what we mean by design and the scope and limitations of its relevance to the management world.

DESCRIBING DESIGN

The word design means many things to many people. Design includes the work of people from a wide range of disciplines: industrial designers and design engineers, architects, graphic designers, illustrators, environmental designers and all those industry and product-related disciplines like textile design, automobile design, furniture design and many others. It also includes the people who are concerned with systems of all these things. In attempting to describe design we do need to agree about characteristics that all design and designers share, and also which of those characteristics are exclusive to design and designers.

However, we are not concerned with the legitimate extension of the word to include 'ideas'-based disciplines. So in what follows we exclude the designs of economic models, philosophical systems, and so on. These 'ideas' disciplines do share one of the three important characteristics which all designers have in common and which is exclusive to design – a methodology (of which more below). With these abstract usages excluded, a useful and simple definition of design which covers all the various uses described above is 'a plan for an artefact or a system of artefacts'. More particularly it is a working definition which is highly relevant to the management world.

Managers are surrounded by artefacts. In the manufacturing industry they call them products, and design is simply the plan that managers make for their products. However, design is the concern not only of the manufacturing industry, which makes and sells products, but also of retailers who buy products in order to sell them and of service industries who use products in order to provide a service. A banker, for example, would be unable to operate without his data; and that data comes from hardware, which itself has to be designed. Furthermore, every business works in a physical environment that has to be designed. Well-designed buildings, interiors, and physical distribution systems all contribute to the efficient operation of every business. Finally, all businesses need to communicate, and they do so with man-made things – reports, promotional and advertising materials, videos, signs, and a host of other products that must be designed.

All these aspects of design can be measured. For example, design input into products, services, and environments has powerful leverage on gross margin. Thus design pervades the manager's life and work in a measurable way. To gain its full potential benefit, therefore, a manager needs to know how best to use design and how best to understand its contribution.

The problem is that not many managers accept that artefacts dominate their world and need to dominate their thinking. Ask a manager what is of prime importance and his answer will range from profits to people, with products a long way down the list.

This disregard for artefacts, and the need to make plans for them has its origins in 19th-century Western culture and the education system which still supports it.[2] We are taught to value ideas above action, things spiritual above things material, the conceptual above the pragmatic and the logical above the intuitive.

The outcome is a manager with inability to appreciate the importance of 'things', and a view of design which relates it to either a 'God-given mystery' or a compensatory skill for illiterates.

With these definitions and confusion in mind let us now discuss the contribution of design to general management in terms of the four management preoccupations described above. Our first tasks are to look at the contribution that design makes to 'innovation'.

DESIGN AND THE CONTROL OF INNOVATION

We have come to accept that innovation is the life-blood of our society. It is innovation that enriches our imagination, supports the quality of our life, and determines the future of our children. Innovation contributes to the creation of wealth; it provides the bottom-line profit in scientific and

technologically-based businesses. Indeed, innovation is now the main-
spring for most managerial activities, the basis of key investments and
the most highly rewarded management function.

No one has been quicker than the designers in identifying their
particular skills with the innovative process. They seem right to do so.
The associations are there: innovation, creativity, invention, and prob-
lem-solving are the words commonly found in the literature of design.
Designers and innovators are perhaps the same people doing the same
things – particularly in manufacturing industry.

Or are they? Are those innovation-related words quite so closely
identified with each other? Is in fact innovation all good news? Can one
question the value of unqualified innovation and indeed its unqualified
relationship with design? Could it be that design has not much to do with
creativity?

We have defined design as a planning process for artefacts. By that
definition, design becomes the key element in the planning processes
of the business – the plan for the things the enterprise makes, sells, uses,
or communicates with. But design will never be an effective tool if it is
confused with the innovative process. Let us go back to that basket of
words – innovation, invention, research and development, creativity,
product development, problem solving – all of them words used in
business to describe a process of change, directed toward the new. Much
has been written in an attempt to give special and related meaning to
each of these words.[3] In what follows let us use innovation to describe
the whole process, recognize that it is a creative process, and argue that
it is something different to design. It follows, therefore, that design is not
a creative process. Indeed, design only deserves its key role in business
when we appreciate that it is not.

The operational activities of a business enterprise – manufacturing,
marketing, providing a service, for example – cannot remain relevant
(and profitable) unless they continue to change. Even if a business
wishes to stand still, it needs to change in order to maintain its equili-
brium within a shifting environment. Many businesses with a cultural or
a nostalgic element – fashion, textiles, and furniture are obvious ones –
need to know how to change back to the old as well as forward to the

new. But change they must. Clearly, therefore, the innovative processes are often, but not always, a necessary prerequisite to business operations. However, these creative and innovative activities never proceed at the same rate as the operational processes. History shows that creative flowering – the eureka principle – rarely happens as part of the calm and ordered flow of historical events. Indeed, it is often the essence of the creative act that is out of time with society, acting as a catalyst for social change, often as a disruptive and now and then as a destructive force in society. The technical advances in late 14th-century Europe and the industrial revolution in the late 18th century are examples. Our current information technology innovations and the biotechnology innovations that are just upon us also provide examples of the unplanned and disruptive nature of creativity in society.

The same is true in the business world. Innovation does not fit comfortably with the annual profit statements. It can and does disrupt the commitments to stability on which the business is built; carefully planned return on capital investment, or commitment to pensions for long-serving employees, for example. These considerations and many others are stabilizing and even civilizing forces. The recent collapse of many innovative information-technology businesses is a case in point.

Clearly, there must be something that modulates, controls, and encourages the innovative and creative inputs into the business – something that makes innovation meaningful. This something is design, which acts as a thermostat for innovation, responding to the voices and views of customers, employees, capital investments, and all the other factors that constrain, sustain and shape a company's culture as well as its operations. Design is the driving force, not only of change (which is inevitable), but more important, of the rate of change. It feeds innovation into a system in a way that preserves the plan for the bottom line. Planning for product (our definition of design) is in this context arguably the most important business function.

To say that design is not itself a creative process is not to argue that designers are not creative people. The interface described above makes it necessary that they should be so. And in craft-based businesses like silversmithing and pottery, the innovative, the design, and the operational activities of the enterprise are usually done by one person. Even in large organizations, the lines are not clearly drawn. They are artificially separated here in order to identify an importance for design that goes well beyond the creative act.

Once this is understood it is easy for designers and more importantly (as we shall discuss) for design-trained line managers to deal with the key issue of innovation, and its encouragement and control.

DESIGN AND QUALITY

The second of the important management preoccupations we undertook to relate to design is quality and its control. Once again before we can describe the relationship we need to deal with definitions.

It carries an aura of exclusiveness. In the 17th century people talked of 'the quality' – meaning the upper classes: and as late as the 70s Fred Hirsh wrote about 'positional goods'[4] in relating quality to exclusive ownership of goods and services by the same kinds of people. Quality also means high standards of make and finish. It carries a flavour of reliability and value for money plus a flavour of the old-fashioned, and the dowdy. Quality products are perhaps a little boring on the outside, but can be relied on to work. We associate quality not only with Heals, and Harrods, but also with Marks & Spencer. Quality is Rolls Royce now and Rover thirty years ago; we associate it not only with individual craft products, but also with mass production.

Many of these subjective value judgements are clearly contradictory. Marks & Spencer and Harrods are not often included in one category, nor is a well-crafted product reliable in the sense that no two of them are likely to be exactly the same. A definition therefore which overcomes all these subjective considerations, and is of relevance to management, particularly in the manufacturing industry, might run as follows: 'Quality is the extent to which a product meets the specification drawn up for its manufacture; and where the product is mass-produced, consistently meets that specification'. Such a definition not only makes space for subjective considerations, but is also capable of measurement.

How in fact is quality (defined in this way) normally measured, and, given the necessary feedback procedures, controlled? Three ways are normally in use in manufacturing industry.

1. By inspection at the end of the process (either of each product, or by a sampling procedure) which compares the product to the specification.
2. By an attitude among the people concerned in manufacture who, through training and exhortation, place quality at the forefront of their thinking during the manufacturing process, and are constantly referring to specification. Quality circles, and related organization systems, fall into this category.
3. By ensuring that the specification itself is developed in such a way that it becomes very difficult not to meet that specification.

All of these ways of dealing with the control of quality have their place and none is mutually exclusive. Nevertheless the third contributes the most effective route. In effect it shifts the problems of controlling quality to a point in the process before manufacture, the determination of the specification.

Now it is also the case that the last stage of the design activity is the specification;[5] and the determination of specification constitutes the heart of all design activity. It is a difficult and complex task demanding the resolution of conflicts arising from all the management disciplines. From marketing which may require product characteristics which are difficult to make, or from finance which may limit product options or require the cutting of production corners.

No ideal solution is ever achieved. Nevertheless it is generally recognized that designing quality into a product is better than inspecting it out. An appropriate specification is worth more than hours or exhortation or an army of inspectors. Design therefore occupies a key role in the determination of quality, as it does in the control innovation.

DESIGN AND MANAGEMENT

Line management performance is perhaps the most important and permanent of the four 'preoccupations' we chose to link with design.

Our task is to explore the extent to which design can contribute to effective line management performance.

That managers do in fact make use of design skills has been an interesting outcome of some recent research at the London Business School, which has identified a complicated and existing phenomenon called 'silent design'.[6]

What this means (in rough summary) is that, in the organizations studied, a great many people (many of them managers) are engaged in designing who are not designers; that quite often they are not aware that they are designing, and do not necessarily agree that what they do is designing once they are made aware of it. Furthermore the process seems to work; though better in some cases than in others. This interesting proposition is at this stage only an hypothesis. But if it were to be confirmed it would carry some interesting implications for the place of design in the management structure. Most of them are to do with the dismantling of some of the shibboleths that surround the design profession.

Dismantling of this kind happened to many other business-related professions long ago. Accountancy skills are no longer pre-empted by accountants. Line managers need to be equipped with these skills in the same way that they all now use computers for themselves, which ten years ago was relatively rare. Even the most defended professions like medicine and the law have, through the advent of information technology, begun to move some of their processes into more general hands.

If design skills are to be available to line managers, we need to say what those skills are. There are three important design contributions that the manager can make good use of. These are: a care and concern for things, a set of special skills, and a special methodology. Each of them deserves some comment.

We have already touched on the first. A care and concern for things does not come easily to managers in our society. Their concern is based on a theory of organizations in which the manager's task is to achieve the objectives of the enterprise, expressed and measured by profit. While doing this, the manager also needs to provide optimum satisfaction for those connected to the enterprise, such as employees, shareholders and customers.

It is a praiseworthy concern, but the objectives behind it are much more easily achieved if managers have a deep interest in what they make and sell. Design training enables us to correct this imbalance – to learn, as the Japanese learned many years ago, that without product leadership most businesses lose their competitive edge.

The second attribute of the designer is visual literacy, the ability to see and reproduce what is seen – in other words, to draw, to model, to create visual analogues. Contrary to the folklore, these are not God-given skills. Anyone can learn them, in the same way they learn to read and write. Most designers have developed these skills early and through constant reinforcement have turned them into a specialization. It is certainly the case that the lack of these skills can create great difficulties for managers, who are constantly faced with massive amounts of engineering drawings, architectural plans, factory layout diagrams, work flow charts, and visually-represented planning documents.

Finally, the methodology of a designer is action-based. In dealing with problems, the designer's concern is to find out how before finding out why. The inclination is thus to take incremental and practical steps to a solution. The way of working is a valuable corrective to the methods that have tended to dominate the education of managers. Their training motivates them towards analysis rather than synthesis and to seek general laws based on hypothese before acting.[7] In other words, managers try to find out why before finding out how. Design methods are a salutary corrective to a view of the world that puts thinking above doing. Design is what management ought to be – 'action-based'.

DESIGN AND CORPORATE STRATEGY

Our fourth and final area of management preoccupation is concerned with the extent to which design can contribute to corporate strategy of the business, or indeed any kind of corporate body.

This, the most complex of the four areas, has been described as making corporate strategy visible: that is, using the powerful visual tools of the designer described above to effect corporate change.[8]

All corporations are constantly under change, whether internally in terms of their organization and their employees; or externally in terms of their markets, and their commitments to shareholders, suppliers, local communities and others.

Corporate identity design which is the usual way of describing this kind of design work is as much concerned with internal as external change. In the nature of things it is usually undertaken by external consultants. The corporate identity designers do not work alone and they were certainly not the first to work in the field of corporate strategy. The very highest level management consultants, those who deal with organizational change were there first. Corporate design consultants have been able to elbow in alongside the occupants of this exclusive club only because they possess skills and tools that the traditional consultants do not – the skills to see and to make others see.

To see, and to help others to see, is an obvious need in identifying the corporation, and helping it to express its people and purposes. It is perhaps less obvious that these observational skills are also the tools by which the corporate identity designer becomes the prime mover in implementing organizational change.

Does this proposition work in practice? Consider what can happen when a management consultant is used but 'seeing' skills are not applied.

Typically, a corporation might want to adjust the management reporting system of a newly-acquired subsidiary to match the system used by the rest of the group. The chairman of the subsidiary accepts the idea that management consultants should be brought in.

The consultants go in, work, report and leave. Then month after month, the management information flows in, impeccable in format, accurate in every detail. Consultancy has 'worked', but will everyone live happily ever after?

One day, a letter from the new unit to the centre requesting some capital expenditure makes a cash-flow forecast inconsistent with management reporting. It turns out that the subsidiary is actually still using a different reporting language. Consulting has not worked. The subsidiary is not, and never was, organizationally integrated.

No one was practising observational skills. For had the group's chairman said that all the vans or the letter heading, or indeed the people, at head office should be painted blue and that a particular operating unit of that division should be blue with green dots and dashes, then at least everyone would have noticed it. The problem of organizational integration would have been no easier by way of a corporate identity route than by a financial control route. It might even have been harder. But it is much more difficult to avoid; and success or failure would have been open and noticed.

The kind of questions which the corporate identity consultant must have answered go much less easily into a pending tray than do, say, the questions asked when the consultancy concerns job grading and evaluation, or the consolidation issues of the annual financial accounts. Most conceptual effort can be stored in written reports, computers or books. Design effort is an implementary activity; it does not really exist until it is completed as planned.

The strength of corporate identity is that of a double-edged sword. By its visibility, it forces decision-making into the open; and by its implementary characteristics, it forces management into a decision-making mode.

This was recently made evident in one major British engineering company with large factories with widely differing corporate identities, one of which, based in the Midlands, was due for closure. The group headquarters, anxious not to identify the company too closely with failure in a part of Britain where it was well represented, was dragging its feet in facing the issues.

Meanwhile a corporate identity programme was well in hand. So the reluctance to identify with the plant that faced closure was made visible by its corporate identity implications: out of all the group's installations, this one alone would not carry the new uniform. The dilatory behaviour of the group management became painfully evident; so the issue was finally faced.

Perhaps the designer should have left the plant alone. But after all, the decision-making was up to the managers; and they could not make decisions until they had seen the problems, but they could not see them until a designer had painted them in bright colours.

All this had little to do with the aesthetic and creative skills of the designers. These may be an additional bonus to a corporate identity scheme, but they are never central to it. Their importance is chiefly this: aesthetic and visual creativity demands high observational skills from the people who practise it; and high observational skills are the strongest management tools in identifying organizational behaviour and implementing organizational change.

CONCLUSION

What therefore emerges from these four contributions from design to the management world is that, while design training can help line managers to contribute to a better design of their products, it can also help them to manage better.

The additional contribution is the most important one. Design, we have already agreed, is a much degraded word, used to mean a basis for the philosophical and economic life of a country as well as a label on a pair of jeans. Designers are stereotyped as back room boffins in white coats or as creative directors in bomber jackets with a two-day growth of beard. Design can make its loudest noises about trivial and ephemeral consumer products and says little about the insides of nuclear reactors or aeroplane engines in which enormous commitments to capital and to culture are involved. Designers rarely attempt to quantify their contributions, preferring to be judged by creativity (which by definition design is not) and by subjective and aesthetic considerations which cloud that contribution.

If we are to ensure that the place of design in the management hierarchy remains secure at least for this decade, then we must sweep away the detritus and distortions which surround it. We have tried to do that in this paper by exploring an appropriate role for design in the management world. We have defined design as a plan for an artefact whether that artefact is a product trade and sold (or bought and sold), or part of the business environment or the information systems that the business uses. We have sought to define a key role for design in four important management fields: innovation, quality and its control, the development of line managers, and the contribution of design to corporate strategy.

However, work still needs to be done to identify and quantify how design can contribute to many more management fields than the four we have been able to cover in this paper. Such work is needed if we are to confirm the most important of our propositions: that the design contribution to management is to enable managers to manage better; that in their hands it is a powerful corporate weapon.

REFERENCES

1. *Policy and Priorities for Design*, Design Council Strategy Group, Design Council, London, October 1984.
2. Weiner, M. J. *English Culture and the Decline of the Industrial Spirit (1850–1980)*, Cambridge University Press, 1980.
3. Storr, A. *The Dynamics of Creation*, Secker and Warburg, London, 1972.
4. Hirsch, F. *The Social Limits to Growth*, Routledge and Kegan Paul, 1977.
5. Gorb, P. 'The Business of Design Management', *Design Studies*, Vol V No 2, April 1986.
6. Gorb, P. and Dumas, A. 'Silent Design', *Design Studies*, Vol 8 No 3, July 1987.
7. Kuhn, T. S. *The Structure of Scientific Revolutions*, University of Chicago Press, Chicago, 1976.
8. Olins, W., *The Corporate Personality*, Design Council, London, 1978.

PART II

CASES

*Sir Terence Beckett spent
most of his career with Ford,
becoming its Managing
Director in 1974 and
Chairman from 1976 to
1980. In 1980 he became
Director General of the CBI,
a post he held until 1987. He
was Chairman of the
Governing Body of the
London Business School from
1979 to 1986. He is a
member of the Top Salaries
Review Body and a Director
of the CEGB.*

Design, Product Planning and Prosperity

My theme is this: design should be practised, not as an art in itself, aesthetically and intellectually fascinating though that can be, but as part of business activity as a whole and product planning in particular, if it is to fulfil its proper role in creating our future prosperity.

At first glance this seems unexceptionable, but from all my meetings with industrialists in this country in recent years, I am not at all convinced that enough of them really do understand the importance of design in product planning; nor do some designers.

The complaint made by many talented designers in this country that their work is not valued or used by industry is matched by a perception that outstanding designs have been associated with spectacular commercial disasters, such as the Mini, Concorde and Rolls Royce Aero Engines. We cannot afford, as only a very moderately successful country, this kind of extravagant failure, nor, on the other hand, can we afford not to use design effectively in our future products.

Let me give it some edge: design like fire, or self-pride, can be a good servant, but a bad master. But lest I be misunderstood, I also maintain that good design, the objectives for which are derived from thorough product planning, will be absolutely essential if we are to turn round the future prosperity of this country. The purpose of this paper is, therefore, to explore this relationship between design and product planning to achieve commercial success and national prosperity.

Let us begin with prosperity. Norman Macrae, Deputy Editor of *The Economist*, writing at the end of last year on how he saw the future, recalled an article written by Keynes in 1928. Keynes said then that he thought the standard of living in Western Europe and America within a hundred years 'will be between four and eight times as much as in 1928'. He went on to say that since nobody could sensibly want to consume four to eight times what they do today, people would come to see the pursuit of money as 'what it is, a somewhat disgusting morbidity, one of those semi-criminal, semi-pathological propensities which one hands over with a shudder to the specialists in mental disease'. This, by the way, from a man who had just made a fortune on the Stock Exchange by judicious speculation. To be fair though, he only wanted to have enough money to enjoy a full life. It wasn't for the money itself

Sixty years on we can see that Keynes is likely to be right in his projections on future growth, but almost certainly wrong in thinking we would not know how to consume four to eight times what we enjoyed sixty years ago. In fact, with a deficit of some £20 billion on our current account this year, we are spending more than we have earned, and most people could spend a great deal more than that, given the opportunity.

But on growth, we are well on the way to meeting Keynes' forecasts. Macrae calculates that sixty years on, if we take population changes as well as inflation out of the comparison, real disposable income per head in the UK has increased by 2.4 times, in the USA 2.9 times, while in Japan it has increased nearly eleven times. However, this growth also reveals an accelerating trend. We have increased our real standard of living per head in this country by 2.4 times in sixty years. It took us a hundred years before that just to double it.

If we leave the Macrae article now and look at future growth, the Austrian economist, Joseph Schumpeter, a contemporary of Keynes, analysed the changing growth rates of the developed countries from the beginning of each of their industrial revolutions. He concluded that, although many different causal elements brought about growth, there were two factors of predominant importance: new scientific and technological discoveries and the availability of capital. There is no doubt that these were the factors that fuelled the growth of the last sixty years.

What can we assume about the future? We have a great deal more scientific and technological knowledge today than we had forty years ago and capital is much more readily available. We are now moving into a steeper region of the exponential curve of growth than in the last sixty years. Wealth creation is increasing at the rate it is because part of our greater growth, of course, is increasing the base from which future growth is generated. Whenever one discusses future growth these days the immediate resistance one meets is 'This won't happen because of environmental problems'. I believe that these problems will in fact create opportunities, which will contribute to growth.

This second revolution that we are now engaged in will be different from the first. The first used machinery and steam to enhance men's muscles. The second will be to a great extent a revolution in the way we use our intelligence.

Increasingly, if you listen to them, entrepreneurs will tell you they are less concerned with buying and selling conventional factors of production and goods. They still do this, of course, but increasingly they find themselves engaged in the buying and selling of ideas, research, know-how and services; in short, the products of intelligence.

What the Calvinists among the environmentalists do not understand, when they suck their teeth at the mention of the word growth, is that this growth will be different. Even in manufacturing it will not be primarily smokestack growth. But it is going to be much more like a growth in services and knowledge. The environmentalists, like the Malthusians at the beginning of the first Industrial Revolution, are generals fighting the last war, which in an age of change has always been a poor preparation for the next struggle.

No. Growth will be more rapid in the next forty years than it has been in the last sixty and it will not be unduly constrained by environmentalism, but it will be to some extent actually enhanced by it.

Remember Schumpeter's factors that determine growth: scientific and technological discovery and the availability of capital. The scientific and technological base from which we move today is immeasurably stronger than sixty years ago. This is not just in more of the same, such as new fibres, foams, ceramics, superconductors, adhesives, molecular coatings, metallic compounds, plastics and so on.

The transistor was still in the laboratory forty years ago. Today we are in the advanced stages of developing many kinds of far-reaching artificial intelligence. It was only thirty-five years ago that Crick and Watson unravelled the double helix of DNA. Crick today is in charge of the project in America that in just five years time will have mapped the human genome in complete detail. We have the super-colliders that are breaking down the last barriers to understanding the fundamentals of matter. And these are just the peaks in a vast range of new scientific and technological discovery. If the fruits of the Industrial Revolution, the opening up of the West and birth control confounded Malthusianism, in the new age, genetic engineering, for example, will not only enable us to feed the world, but by making the deserts bloom, we have the means to reverse the build-up of carbon dioxide and the greenhouse effect that so exercise the environmentalists today.

But one could go on and on about the vastly stronger base of scientific and technological knowledge we have today compared with sixty years ago.

No-one would doubt, I think, that the second of Schumpeter's prerequisites for growth is present in abundance: the availability of capital. We have an army today of able people spread around the world in our financial centres, amply supplied with funds, packaging and repackaging capital in ever more fertile ways to enable enterprise to have the means to put new science and technology to work.

But what I think we have to try to grasp is not just the potential size of the growth that will come from these developments, but that this will be transmitted by products with shorter development cycles and much shorter lives.

There are some companies today whose market and supplier base are nothing less than the whole world. As new technology makes economies of scale more significant and transport costs fall as a percentage of total cost, this globalization of local markets will increase. With modern communications and intelligence, small improvements anywhere in the world will be eagerly snapped up, but those items that are not quite as good in satisfying customer needs will be much more quickly and decisively rejected and scrapped. Competition will become more intense, with fewer hiding places for those who are not quite up to scratch.

What should we do about it? . . . Let me tell you a story.

It was almost thirty-five years ago that I was put in charge of the product planning function at Ford in this country. Until that time, product development took place when the Managing Director got his chief people together and told them 'I want a new car in two years' time, about the size of the Morris Minor. We shall need four hundred a day. The weight will be 1600 lbs. The engine displacement 800 to 1000 cc. The acceleration will be 0 – 60 in 26 seconds and the touring fuel consumption has to be 40 to the gallon. Now I want some styling and engineering ideas from you and we will meet a month today.'

Sir Patrick Hennessy, who was Deputy Chairman and Managing Director of Ford at that time, realized that with increasing competition, this occasional direction would not be adequate; product development would need full-time continuous attention.

The brief statement of my functions as the General Manager of Product Planning Staff was to improve the competitiveness and profitability of our products. I had a multi-disciplinary team consisting of engineers, cost and finance accountants, marketing, purchasing and manufacturing people – and a formidable set of problems.

The basic difficulty at that time sounds incredible today, but life was not competitive enough. Practically no-one in the company wanted to change anything. Manufacturing regarded my activities in developing new product as downright disruptive. Did I not know that the only way a mass-production manufacturer could succeed was by producing more and more of the same thing, hitting production schedules, getting costs down, improving quality? Changing things all the time was destructive.

We even had the same attitude in Sales. I remember one new

product introduction, in my early days in product planning, where the Sales Manager apologized to the dealers for having to introduce a new truck. He implied it was not needed, because there was lots of demand still for the old one.

My inexperienced team as well was struggling with the immensely complicated task of trying to get their fingers round the relative importance of the different elements that were needed to plan and control the ten thousand bits that go into making a new car or commercial vehicle.

We slowly made progress. Patrick Hennessy believed that product was the backbone of the business and he gave me splendid and invaluable support. Incidentally, if the attitude to product I have described as necessary in the new age is to come about, I think it is essential that the Chief Executive believes and acts in a way to make the customer, product and design of superordinate importance in a company. No-one else can really do it. And my advice to any young man or woman who finds that the boss does not believe in the value of the new product is to leave him and his company, because there will not be much of a future there anyway. As Machiavelli said: 'One cannot help a foolish Prince'.

Sir Patrick formed a Product Committee of our leading executives, including the bosses of Manufacturing and Sales. He held regular meetings to review our product plans, so that they began to see each developing product as theirs, and not just something imposed on them.

The method of analysis and control we used to plan our products with the support of the whole organisation was by means of a product proposal. This began and ended with the customer. The proposal outlined a strategy. It analysed the customer researchon our own and competitive products. It came to conclusions on this. It moved on to describe the specification of a car, to set out its future sales volume, its cost, price, and profit, the timing plan in complete detail, and the development and tooling and equipment costs. The proposal set out the new manufacturing methods and processes that would be used. It described the investment that would be needed, the return you could get on it, the cash flow that this new product would superimpose on the company as a whole, the source and application of funds and a detailed sensitivity analysis of factors that could change and affect the outcome of the project. A comprehensive analysis of probable competitive actions was factored into the analysis, including the likely reactions to our own product introductions. The proposal concluded with our marketing plans to introduce the car to the customer. If you are in the design business and you do not think you could provide your clients with this kind of analysis, could I suggest you consider hiring an MBA from the London Business School. They do two or three studies like this before breakfast every day, while they are working for their MBA.

We gradually refined this analysis and learned from our mistakes. Product staff began to take a more pro-active role: you get nowhere simply putting many different divisions' ideas together. Multiple objectives needed to be consistent, one with another. They had to have tasks in them, but they needed to be feasible. Above all the data needed constant review with the customers' changing requirements reassessed.

What we did not realize then was that these methods, which we were so painfully learning under Patrick Hennessy's guidance, would be absolutely essential to us as the competition became more intense. Nor did we know that, by and large, our British competitors were not carrying out this kind of analysis and planning. Just as one example, which I discovered later, Ford spent as much on market research, in those days, as the rest of the motor industry put together. But nearly all of what the rest of the industry did spend was on existing product; nearly all we spent on research was on future product.

To return to my story: the competition was warming up by the end of the 50s and the Sales Managers no longer thought we could go on selling the same old merchandise for ever. Even the manufacturing people began to realize that if we did not have new product, their plants could go on short time or might even close down. Competition, like democracy, is a painful way of doing things, but is better than any alternative there is, as the Communists are beginning to discover.

1960 was a watershed for me and my team. We had introduced in the previous autumn, as our small car, the new reverse back-light Anglia, but BMC had introduced the Mini in a class below this, a class that no British manufacturer had previously occupied.

The Mini rapidly established itself as the designer's car *par excel-*

lence. Its whole development had been dominated by Issigonis, a designer of genius, who had dreamed for years of developing a car with the ultimate in compactness. It also provided the press with what they had been telling the British motor industry it needed: innovation in design, which they then generally short-handed to 'Design'.

Issigonis did things differently from the way we were doing them at Ford. His product development was autocratic, not that of a team. He pursued one idea, compactness in the case of the Mini, with single-minded dedication. He did not brook criticism or need research.

To achieve compactness in the Mini, Issigonis turned the engine around from north-south to east-west, he used front-wheel drive, and independent suspension all around. The press had something to write about. And, by God, the car was British, unlike those Ford products. This latter ignored the fact that a man called Issigonis was their designer and my design team was absolutely solidly British. The only thing the press did agree with us on was that our team was solid all right.

When the Mini was introduced, it was priced lower than our lowest priced car, so that for the first time since Ford had introduced the £100 car in the mid-30s we had lost the bottom price slot in the British market. This caused consternation in some Ford hearts.

The Mini became the toast at parties. Duchesses went chauffer-driven in it to Harrods. Some of the Royals even seemed to prefer it to the horse – for a time. And everyone recognized the Mini as a triumph for design.

However, in spite of the praise lavished on the Mini, all was not as it appeared. Even before it had been announced to the public, I had sidetracked one on the way to a dealer, driven it and been all over it, before letting it go on its way. I came to two intriguing initial conclusions: first of all, it did not seem to me that BMC had got its cost/price/profit equation right. Secondly, compactness had been pursued to the detriment of interior space and a real luggage boot, which I knew most customers set store by.

Just as soon as we could buy a Mini, I had the whole car stripped down to its spot welds to be quite sure of its cost. This revealed an extraordinary state of affairs. At the cost we in Ford could have made a

Mini, and we knew we had the edge on BMC on both manufactured and purchase costs, each car was making a substantial loss at the price it was being sold.

Did we deduce from this that BMC were playing a destructive game to buy share at our expense, or were they convinced that they were making a profit at that price? We could not be sure. I had a market research study made of the Mini and established that they could have priced their car £25 – £30 higher, at, or about theAnglia's price, without losing any significant sales.

I inclined to the view that they did not know their costs and also that they supposed that a Mini car, in a size class below the Anglia, had to be priced below it. This ignored the fact that a watch can cost more than a clock. It is the number and complexity of parts that chiefly determine the cost of a car and not just kilograms of material. And the Mini was a more complicated and more expensive car than the Anglia.

We were faced strategically with two problems: one, should we follow BMC with a car in the Mini class to avoid losing share and, two how were we to respond to BMC's prospective 1100 car, a big brother to the Mini, which had all the innovation of the Mini and would presumably not have the same restricted interior that the Mini had to offer in its extreme pursuit of compactness?

I was getting all kinds of useful advice. I was told by people in the company, by dealers and visiting management from North America, looking at the popular acclaim for the Mini, that, 'Surely anything those BMC guys can do, Terry, we can do better?' What they meant was: 'Why don't we copy them?'

I had to say that if we did what BMC were doing, according to my calculations, we would join them in a disaster. My colleagues could not believe BMC were making an error of the magnitude I described to them. And I had to admit that I could not be sure. The French Existentialists, as you know, attach a particular meaning to the word *angoisse* in describing the human condition where you have to commit yourself to a course of action, without being sure that you are right. I had more *angoisse* at that time than I could usefully handle.

Had we missed something out of our calculations? Was there some other factor we needed to take into account? If we concluded we were right about BMC, what on earth could we do about it?

I was clear on one thing: there was not any way we could follow the Mini. BMC were not only making a substantial loss on every care they produced and sterilizing a vast piece of their assets in the process, but the further, probably unintended, but diabolical consequence was that they had denied that segment of the market to the rest of the industry.

The besetting worry was would the Mini roll up our left flank on the price scale? As 1959 moved into 1960 we gradually began to breathe again. It was quite clear that the roomier interior of the Anglia and its boot meant that many customers preferred our car to the Mini in spite of innovation and price. In fact, in the first year of production of both cars, nearly twice as many Anglias were sold in this country and abroad as Minis. And we were making a profit on every car we sold; not a large one, this is difficult on small cars, but significant, and multiplied by our greater-than-expected volume, it gave a better-than-planned return.

This, however, was only part of the problem. What were we going to do about the Mini's big brother?

We had planned to introduce the Classic car in 1960, twelve months after the Anglia, in the medium-car segment between the Anglia and the Consul. It was delayed because it was to share its engine with the Anglia, but the Anglia was doing so well we decided to postpone the Classic until we could break the engine bottle-neck. But while we delayed the introduction of the Classic, it became clear to us that if we were to meet the challenge of the BMC 1100, we needed a different strategy.

We would still introduce the Classic as the next car up, but we knew it was too small, too heavy and too expensive for the new competitive challenge we now faced. In the rather desperate military metaphors we used at the time to cheer ourselves up, we said our left flank on the price scale was pinned down by the Mini, and so we had to advance our centre. Every amateur strategist will tell you this is fraught with danger. We had lost our claim to have the lowest-priced car. We now had to have a product which would give better value for money than the competition, further up the price scale. And this germ of an idea became the Cortina.

We estimated the size of the BMC 1100, by a triangulation of power to weight calculations, knowing the potential of their engine range and projecting body weights to derive a likely package. We made sure our car was comfortably larger than this and, in fact, would provide an interior package to draw business from a whole raft of medium cars such as the Hillman Minx, because, although its size was comparable, it would be considerably lower priced.

Our undramatic concept was good interior roominess (including a large boot, which neither the Mini nor the 1100 had) for a lower price than any other car of this size had had before. If the opinion of some people I discussed these proposals with was not encouraging, the research we conducted was, provided, and it was a necessary condition, we met our cost objectives. If we did not, the Cortina would simply be another medium-class car, with mediocre volume and profit and no answer to our problem.

There is one thing worse than making a mistake and that is to continue with it. The new strategy to introduce the Cortina meant the Classic was doomed to die young. The Classic had been a serious mistake on my part. There was no hope of recovering its costs in the short life now planned for it. Development and tooling costs on the Cortina had to be added to those of the Classic and we had also embarked on a heavy programme of commercial vehicles at the same time.

Taking these together they were making a hole in our profits. I should have had more foresight. After all, I did have a ten-year forward programme for our products and a shadow programme for each of our chief competitors, but I had not taken the probable impact of the BMC 1100 seriously enough, early enough.

I was not yet out of the wood on the plans for the Cortina. If it was to succeed, it had to draw business away from the medium-car sector. During a presentation in the USA on our plans and the marketing staff man said to me, 'But you don't have a market for this car'. Strictly speaking, in terms of the then current market analysis, we did not. When I told him we had got to make a market for it, I think he suspected me of smoking opium. But, if you are going to do something new, there may not be an existing market for the product you are planning. In fact these days, you ought almost to beware if there is one. But, on the other hand, your view of the future must be realistic and it isn't easy.

We needed to act quickly in developing the Cortina; BMC had at least an eighteen-month lead on us. Front-wheel drive was out as far as we were concerned. We had experimented with it at our research establishment at Birmingham for five years and had not satisfactorily cracked the final drive problem. Twenty years later it was different, but at that time we did not have a satisfactory design.

I am convinced it is fatal in planning a product to adopt an unproven design in the hope you are going to get it right in time for production. This, with inadequate cost control, caused what was perhaps the most prestigious company in British industry to crash: Rolls Royce, with its aero-engine development.

Incidentally, Ford of Germany and Ford US were developing a very similar car to the Cortina at this time under the code-name Cardinal. But they decided to go for front-wheel drive and this was the first time either company had used this system. The car was not a success, though there were cost and quality problems as well as those from front-wheel drive. Ford US decided not to market the car in the USA, just ninety days before job one, and the car, sold in Germany as the 12M, never achieved its sales or profit objectives.

We decided that in the Cortina we would follow a conventional layout, with the engine at the front and the drive at the rear. We had a good independent front suspension, using Macpherson struts, but we would have to have semi-elliptic springs at the rear. We knew the press would tell us once again that these would be 'those same old cart springs'.

However, here the story takes a turn, because I have implied so far that all the good design was in BMC, although it was wrongly directed.

It was interesting that the semi-elliptic springs in our rear suspension gave quite a good ride and, because we were not screwed down by compactness, the wheels on our car could move through a greater vertical distance than the Mini's. Independent all-round suspension on the Mini and the 1100 was an interesting concept, but you could travel over bigger pumps in the road in our car without bottoming. General concept in design and actual performance can be clean different things.

We decided that the Cortina would have the outstanding engine from the 940/1300cc oversquare range that in souped-up forms was winning rallies all over the world in specialist cars. It would also have the gearbox from the Anglia, which was superb for the silkiness of its change and its great reliability. Again, our gearshift was simple and direct. The Mini's and the 1100's, with their remote indirect linkage to the front of the car, had awkward lost motion.

To achieve a kerb weight target of 1625 lbs we needed a body which would combine a roomy interior package and a good-sized luggage boot, with the strength of much heavier cars. Our engineers in America had just made a breakthrough in body floor design for integral construction cars that provided just this. This together with a number of most useful new methods developed by our own body engineers would enable us to get the right torsional rigidity for the body at a low weight.

We went through the whole product planning procedure I described earlier, and I do not have the space here to describe the detail in which it was done, but it was thorough. Some of our earlier product planning failed because we had used too broad a brush and hoped that it would all come out right in the end; it did not. This time we addressed all the relevant detail.

There is one more relationship between product planning and design that I should like to draw out in this paper. Some people claim that in giving detailed objectives to your designers you stifle their creativity. My experience is exactly the opposite. We gave our stylists nine closely typed pages of objectives for the Cortina, as well as package drawings whose hard points had to be met, and we achieved an appearance which was not only outstanding them, but twenty-five years later is still outstanding.

The skill of the designer, the genius of the artist, is to observe the limitations of his medium and the objectives we give him, but to transcend them. As I can only use words here, let me transpose into them. The product planner would say he wants a sonnet of fourteen lines of iambic pentameter, consisting of three quatrains and a couplet and the subject should be the brevity of life, the fickleness of fortune, or, more encouragingly, my lady's eyes. But the mystery of creation is that emerging from this chrysalis of rules, poetry can take wing. Let us be clear, this act of creation is something that cannot be commanded by objectives, but in the process of achieving it, we can require our objectives to be met.

Back to earth again. We took just nineteen-and-a-half months from clay model approval to job one on the Cortina. We had never done better than twenty-eight months before. In spite of starting eighteen months behind the BMC 1100, we introduced our car at the same time they did. Because we knew we had to do every single thing right, first time, and there was no going back, like hanging, it concentrated the mind wonderfully.

We hit our piece cost target with about one hundred and eighty old-style pence to spare. On kerb weight we were right on the button at 1625 lbs. Our development and tooling costs were under-run by nearly £200,000. We in Product Planning knew that a carefully thought-out product had achieved every objective we had set ourselves, and we had done the job we set out to do. Our latest research showed we had a winner, but the rest of the world was still unimpressed.

When we introduced by the car it was panned by the press for lack of innovation in design. This criticism was mitigated a little by the fact that I had involved Colin Chapman of Lotus to develop a Cortina Lotus during the final months of the development of our car. This was a diversionary tactic on my part, to draw the fire from our conventional engineering, but I also wanted it to illustrate the upward series potential of the car and to sprinkle a little star-dust on it. At the introductory lunch to the press, Colin quite modestly described what he had had to do to turn our goose into his swan and the journalists roared their understanding of his difficulties.

The press gave disproportionate attention to the Lotus connection I was glad to see, but the criticisms of our conventional engineering remained. The BMC 1100, which was introduced at the same time as the Cortina, had all the innovations of the Mini plus interconnected independent suspension. Our old cart springs were decisive as far as the media were concerned. 'Design' triumphed.

I had similar problems with our dealers and even our own sales staff. I addressed meeting after meeting to make sure they understood that, although they had lost the lowest priced car, what we were about to offer the customer was better value for money. That was not the easiest idea to sell. The roomier package and the boot that the family man and the travelling rep so badly needed, a first-rate engine and gearbox and an effective suspension seemed unimportant compared with that magic word innovation. I think they admired my presentational persistence, but they were not convinced. Design, as I said, was riding high.

However, the customer knew what he wanted. In twelve months, the Cortina outsold the 1100 by more than 2.5:1. We produced 264,000 of them in the first year, an all-time record for the first year's production of a new model, not only achieving outright sales leadership at home, but simultaneously making it Britain's leading export car. One in four of cars of all makes exported from this country in that first year was a Cortina. Many journalists and dealers then told me that they always knew it would be a great success. It was nice not to have to argue with them. The Cortina was to remain the outright sales leader for the next twenty years. What is more it was a highly profitable car, combining good unit profits with high volume.

However, to come to the point of my story: why did this approach work? We had separated signal from noise on what the customer really wanted. We had designed to satisfy the customer; our chief competitor had designed primarily to satisfy a design. Innovation is essential, but as Quiller-Couch said on the art of writing, what distinguishes good crafts-manship from the rest is whether it is appropriate. Our innovation provided the customer with what he wanted. We had recognized and digested our mistakes with the Classic. We had worked without our limitations; we were not prepared to offer an unproved design to the customer; the design that we used was, in fact, outstandingly good and what the customer wanted. We knew what we were doing on cost, price and profit, because this is essential if you want to prosper or even stay in business. In sum: our approach to design through product planning was superior to design on its own.

Rover, today's successor to British Leyland and BMC, is a professional company with some good products. Working with its Japanese collaborators and its parent company, British Aerospace, it can have a successful future. But it has lost about two-thirds of the share of market its constituent companies commanded thirty years ago. By contrast, the profitability of the Cortina car line was one of the chief reasons Ford in this country was able to grow and prosper. The dramatic divergence in the two companies' fortunes over the years, in spite of enormous government assistance to BMC and its successors, was attributable to some considerable extent to the resources they had available. This in turn depended on their profitability and ultimately on their different approaches to product planning and design thirty years ago.

These lessons are relevant today. As I tried to show you at the beginning of this paper, the most striking characteristic of the new era we are now entering will be the sheer pace of change. And change will be transmitted by new product. New product will mean we shall need more effective design, but that design must be pursued within the context of bold, comprehensive product planning, based not on what the customer needs, but on what we think he will need in the future.

To conclude: what do we have to do to make this whole process more effective? Let me suggest three things. The first is a better general recognition of the importance of design, within the context of effective product planning. The second is a wider acceptance by companies and design consultancies of the need to create multi-disiplinary product-planning teams able to grapple comprehensively with the opportunities of the new age. In particular we must get a few products for this country that are world leaders. We do not have many of them, you know. And the third, to provide a focus, should we form a Product Planning Society to promote the skills that are needed? We have societies for marketing, purchase, finance, engineering and design, but none for the activity that puts them together to enable them to succeed: product planning.

Jane Priestman is Director of Design, Architecture and the Environment at British Rail. Prior to that she held a similar position at British Airports Authority. She is an Honorary Fellow of the Royal Institute of British Architects and has wide experience in all aspects of design.

Managing Design at British Rail

British Rail is undergoing a major cultural revolution. Yes a real revolution. In line with other major organizations, it is learning that design is a resource that can no longer be squandered.

A radical new system of management is being introduced to emphasize the commercial nature of the business which is demanding effort and dedication on the part of everyone involved. Most of all, it demands total commitment on the part of the Main Board – a commitment to raising overall design quality, not as surface treatment to prop up our ailing image, but as a fundamental tool for management; as fundamental as finance and personnel. Like them it is a business requirement.

Like them also it needs a clear central policy, and like them it must be managed at the start of every project, every development, because it affects every decision made, and how these decisions are to be perceived by the public.

The public judges the management of a business by the visual display on offer. Hour by hour, day by day, year by year, our customers judge the intent and quality of British Rail management through what they see and experience. They know what they are buying and they require value for money.

If the customers are to be happy then design needs to be effectively managed, and to make this happen it needs to be built on a firm foundation. In British Rail I believe there are five foundation stones for the effective management of design.

The acceptance by the board of the role design can play is a commitment. I see this as the first of the five foundation stones of design management.

The second foundation stone is profitability. This is the proof that design is not just somebody's subjective idea of good taste, applied after projects have been fully formulated, but is a key foundation stone for success, built in from the beginning, to improve quality and reduce costs.

The third foundation stone is co-ordination. This is as important to small-scale operations as to large ones. Successful co-ordination allows each activity to work well with its neighbour without stifling the freedom associated with individuality, and without territorial tugs-of-war which are such a drain on energy and innovation.

The fourth foundation stone is perspective. It is not enough for any manager to see only his part of a project – and this is particularly difficult in a large public enterprise. Departmental managers often get too involved in basic administration and in seeing that their own sectional interests are not left out of the pecking order, than they are not able to see the wider issues. To avoid self-absorption every manager must be aware of the contribution made to the whole – and also of exactly where the goal-posts are.

The fifth foundation stone is understanding. It is necessary for any manager to have more than strong personality and conviction. This is not enough to persuade the disbelievers. It is necessary to have a common-sense approach which takes into consideration the problems faced by groups of specialists – who often have very little in common. Only then can bridges be built and often conflicting forces be pulled together.

However, when considering the role of design it is not enough for a design manager to have these five foundation stones, or prerequisites. They cannot operate in isolation. It is vital that the importance of design – and the contribution this makes to profitability – is understood throughout the company; and that means effective training and education.

Thus the design manager in any business must be co-ordinator, motivator, communicator, interpreter and, most important of all, catalyst. In British Rail it is necessary to be something of a juggler, as the organization is particularly complex and the task involves creating a corporate personality which is instantly recognizable and yet at the same time allows for wide retional individuality. The last thing we want is to impose an impersonal sameness throughout the country. Apart from being undesirable, such an attitude would also be impossible to implement. British Rail is not just one business now, but five.

THE BRITISH RAIL BUSINESSES

The first is InterCity, the flagship sector, epitomized by the high-speed train providing a network of fast, frequent and high-quality service between major towns and cities. It is the sector most in competition with the car, the plane and, increasingly, the motorway coach. Investment totalling £600 million on rolling stock, electrification and new locomotives will be undertaken between now and 1991. Real growth has been achieved during the last four years and much of the new business has been at the quality end of the market with the reintroduction of Pullman style and quality. It is searching for new ways to become more competitive.

The second business is Network South East. The largest of the five, it has an annual income of more than £700 million which is earned from its business of carrying 1.3–1.4 million customers every working day in and around London. The business operates rail services from 900 stations – from Weymouth and Salisbury in the south-west, to King's Lynn in the north-east of London.

Its new high profile is supported by a major programme of reinvestment in new trains, track and signalling. This also includes station schemes from new rural stations to vast development schemes such as Liverpool Street and Broadgate in the City of London.

The third sector is Provincial, bringing together the remainder of the passenger rail network, with cross-country services between cities not directly connected by InterCity. It also provides rural services in Scotland, Wales and the more remote parts of England such as Anglia, Cumbria and Cornwall.

Needless to say, Provincial has the least earning power in the passenger sector as it is responsible for the bulk of the nation's railway stations, many of which are less than suited to modern travel.

Out of 1500 stations, 700 are unmanned. The concentration to date has been on providing improved rolling stock. Now we can start a comprehensive programme on station facilities.

The fourth sector is Rail Freight which provides tailor-made transport and distribution services to the nation, from pet food to gravel and from cars to power station coal. After the punishing consequences of the 1983 miners' strike, Rail Freight has gradually re-established itself as a viable force in transport against tough competition from the motorway network. In 1988, for the first time, Rail Freight carried hot edible oil in purpose-built tankers. Here efficiency in design and delivery go hand-in-hand as any delays en route would mean the oil solidifying.

The fifth business is Parcels, with spectacular success in the parcels service Red Star. It now provides customers with air-linked services to remote parts of the UK, Europe and North America. The UK network of 600 outlets controlled by a computerized management system is easily identified through branded Parcels Points.

Design in all five businesses has an effect on how the public and staff perceives the parent company – not just visually, but how good the service is, now clean the trains and stations are, whether things work. All this adds up to how good the management is and this includes the areas which are not normally seen by the passengers – such as the working conditions for the employees. For without the understanding, support and enthusiasm of the people who work for the organization, long-term success will not follow. In our case they affect not only British

Rail's relations with passengers, but also the way these five different businesses relate to each other.

In order to make this corporate stance quite clear, British Rail has brought three of its major concerns – architecture, design and environment – under one directorate. Until last year they were under three separate controls. It is my responsibility to bring these three disciplines together and this is a major step towards a co-ordinated design approach for BR and therefore towards a new perception of the business by the public.

It would be impossible to undertake such a mammoth task without establishing structures to ensure success. Firstly, a design policy statement has been agreed by the board as part of its corporate rail plan. Secondly a Design Policy Committee has been established, chaired by the board's vice-chairman. This acts as the executive arm for change and makes long-term decisions about future developments. It is also necessary and important to take note of external views, so two advisory design and environment panels breathe outside expertise into the system. So, we now have a new structure; we know our role. What is the value of our input?

To be effective in design management, as in any other area of management, start with basics – the analysis of the problem. Let us start with the things the public sees – the evidence of change.

Architecture

This, of course, involves the work of engineers and interior designers as well as of architects. It concerns not only the 2500 stations, but depots, workshops and office buildings. 900 projects, currently in hand, from major commercial developments such as Broadgate, Liverpool Street, and King's Cross/St Pancras to listed Victorian station conversions and unmanned halts. Stansted and the Channel Tunnel terminals offer exciting possibilities.

We have a unique heritage. There are 700 Grade I listed buildings in the process of being adapted to a constantly changing business and an increasingly demanding public. This is a considerable design challenge – and requires determined, strong design direction and control. Our intention to employ increasing numbers of external consultants demands new management skills.

We must fit modern shopping malls into Victorian stations. We must decide whether to refurbish fleets of rolling stock; whether to spend £4 million on restoring listed Brunel trainshed roofs at Paddington, which most people will not notice apart from the fact that the roof will be a little lighter and will not leak; or improving amenities like car parks, eating areas and waiting rooms. We are aiming for lighter, easily-supervised spaces which will be enjoyed. At main terminals we will offer special comfort in Pullman lounges.

Obviously, the aim eventually is to do all these things. But when choices have to be made, for maintenance, safety or for financial reasons – and those reasons are not always obvious, positive communication is vital, both to the staff and to the public.

Industrial design

This small department covers the development of new rolling stock, ticketing and station amenities. We work closely with external consultants and the business sectors to improve standards, and to advise on and contribute to all new industrial design projects from large ones like new trains to the components for a new desk system for our ticket and travel centres.

The Graphic Design Department controls British Rail's corporate identity, its classic double arrow symbol and rail alphabet. In addition, liveries and literature for the five businesses, although often designed by outside consultants, must be constantly controlled, monitored and updated to keep pace with business requirements.

Environment

British Rail has a major commitment to the national environment. For example, we own seventy major viaducts. Twenty-nine in Scotland and twenty-one in England and Wales are redundant. The public is passionate about viaducts and the local pressure to retain them is understandable. Gradually, where possible, BR is finding ways of handing over to local communities so that their maintenance can be assured.

Sensible urban landscaping is vital. There is significantly less vandalism in areas which are attractive and pleasant to be in and design plays a positive role in the use of unbreakable glass and other vandal-proof materials, but most particularly in lighting. Visible, well-lit stations mean that there are not places for people to lurk and no hiding places where they do damage.

The Corporate Environment Fund (£2 million in 1988) matches pound for pound the value of contributions made by other bodies, or by the local community to improve the railway environment. This includes art.

We have, for example, run a competition for artwork to be installed in the new Reading station. Alexander Belschenko was selected. We have installed sculptures in Brixton and Wakefield where Charles Quick's 'light wave' was dedicated in January 1988. Lighting at night is synchronized with train movements. In each case we have sought partnership with property or industrial interests, and worked closely with local public arts groups.

At Lime Street Station in Liverpool, we worked directly with the Tate in the North (in Albert Dock) on a major urban sculpture project to coincide with the opening of their gallery in May 1988. Having provided the amenities which the public expects, there is still a job to be done in raising people's awareness of the efforts being made on their behalf and to do this we believe in involving them as much as possible in the improvement of their environment. Our exhibition – 'Platform for Artists' – toured the country in late 1987.

Finding new uses for non-operational British Rail land is one of the responsibilities of the BR Property Board. Our old trainshed at Manchester Central is now used as an exhibition and conference centre. Railway arches have converted to become central to inner city development and provide stylish commercial outlets, or low-cost start-up units for small businesses. The wealth of BR-owned bridges offers great opportunities for colour and visual excitement and they are often seen as the precursor to economic regeneration.

THE STAFF AND THE PUBLIC

The aspect of design that the public does not see is the education programme throughout the organization. However, accepting design as a management tool, it follows that education equals understanding, understanding equals enthusiasm, enthusiasm communicates to the public and public perception can become more positive.

I believe that it is necessary to have close contacts with the personnel director to develop procedures that encourage staff at all levels in the understanding of the relationship of design to their business performance. While any manager of the organization has a negative attitude, the effect will be a negative response from the public. Of course none of this has any relevance if the staff do not understand the message.

We also must be aware all the time that we have a greater pressure of passengers, the fastest growing network in Europe and the greatest number of old stations. Every architect designer is ambitious to improve, but has physical and logistical problems to cope with. These do not start at the drawing board, they start with the board of directors.

Luckily in the case of BR, there is a great groundswell for change: but it must be directed and managed steadily, and that is our key task.

British Rail is one of the most noticeable service industries in the country – highly visible to the public, as everything is seen through a goldfish bowl – and also sharply noticed by government and to industry. We are, therefore, in a position to give strong leadership in British design, to show the faith we have in the expanding role of design within, to prove the value of design in terms of aesthetics and efficiency and to show the contribution it can make to profitability as well as to pleasure.

I have an obligation in all these areas. The doors are wide open for British Rail. There is a turnround in attitudes within the industry – from the enclosed, defensive and introvertive management struggle of yesterday, to one that is more receptive, responsive and outward-looking. Habits of mind are beginning to change with our cultural revolution and we are beginning to see a very bright future.

Raymond Turner is
Managing Director of Wolff
Olins, a leading corporate
identity design consultancy.
Prior to that he was Design
Director of London Regional
Transport and has also run
the design consultancy arm
of Kilkenny, which was the
body responsible for the
national promotion of design
in Ireland.

Design into Management

INTRODUCTION

This paper deals with the practical steps which have been taken by London Regional Transport over the past two or three years to establish design as a corporate resource, and to show how that resource is being structured to give managers freedom of interpretation on the one hand, while providing co-ordinating guidance on the other.

I have been involved in the process as design director for most of that period, and during the twenty years I have been working as a design professional there has been a significant change in the status of design. It has changed from being essentially a practical jobbing tool to being something else as well. Something which can provide a major planning contribution to company development. Something which in some small way can help a company move from where it is now to where it might be in the future. However, the business world is faced with a problem, a problem of an incomplete proposition. Let me try and explain.

DESIGN AND MANAGEMENT

At one end of the spectrum of design activity we have design – the process itself. As a result of clever promotion, good salesmanship, or simply waking up to the fact that design pays, that it leads to increased sales, lower production costs, a few more percentage points of market share, the ability to stay in the market place a little longer, or even leads to making a lodgement in the bank; the business world is beginning to accept that there is value in employing the skills of a designer in the survival stakes of today's commercial environment.

At the other end of the spectrum of design activity we have management. They have been listening to, and have been persuaded by, the design evangelists who have been saying that you must manage your design affairs as efficiently and professionally as your other corporate resources, like finance or personnel.

What we have are the two extremes: design is a proven useful resource (and the knowledge that managing that resource is crucial to its effectiveness). What we do not have is a significant track record of experience in how to do it.

The major problem associated with design management is not managing design projects (that, like general project management, is straightforward), it is not managing a design studio (that is like managing any other professional office), it is not managing designers (they require no more personnel skills than other specialist groups) . . . the major problem to overcome in making design an effective business tool is how to put design into management.

That is what I want to talk about today – or more specifically how London Regional Transport is putting design into management, so that the design resource can be used:

- at the right time
- in the right way
- on the right job
- for the right cost
- to bring the right benefit.

A NATIONAL INSTITUTION – LONDON'S BIGGEST BUSINESS

In order to understand what is needed by the corporation today it is important to understand how the company developed, what part design played in its historical development, and why the demands on design today are, ironically, diametrically opposite to those of fifty years ago.

When any consultant becomes engaged by an organization, and the same applies for that matter to anyone like me, for instance, whoever takes up a new post in a company, they must become totally familiar with all aspects of the organization, both internal aspects and external. That is no less true for design; and in my particular case doubly important because I was employed to establish a new corporate function.

The one thing I soon discovered was that London Regional Transport is a little unusual, when compared to a lot of other companies, in perhaps three ways.

1. The size of the organization – it is big.
2. The enormous difference in scale of some of the design projects from a 5p brochure to a £200-million order for new trains.
3. The scope, and even spread, of design work across all three areas of design focus – environmental, product, and information design.

I make no apology for the history lesson that is to follow, because it is important to understand a little of the past to see the relevance of what we are doing in the present, and what we are planning for the future.

London Regional Transport is responsible for organizing transport for London, and providing the bus and underground services. We operate in one of the biggest cities in the world in terms of population and geographical spread. Four out of every five people travelling in and out of London during the rush hour travel by public transport.

Some other statistics may be useful in understanding the organization:

• it has the oldest underground system in the world dating back to the 1860s;
• much of it is deep tunneled presenting enormous engineering problems;
• the system carries 5.5 million people every day;

- it has an income of £1 million per year;
- it employs 48,000 people;
- it owns 250 railway stations, 17,000 bus stops, 450 trains and 5000 buses;
- it runs its own power station;
- it boasts having the six longest tunnels in the world.

We have all heard the 'motherhood is good' statement that 'design is important to any organization', but in fact London Transport was one of the first corporate bodies in Britain to recognize this, and to use it as a resource. However, that was over fifty years ago, and our story starts long before that, in the reign of Queen Victoria.

An historical profile

The Victorian era left public transport with an inheritance. Beneath the ground the Metropolitan Railway, opened in 1863, was the first underground railway in the world, steam-hauled by locomotives with special condensing apparatus to get rid of the smoke. Overground, public transport relied on horse-drawn buses on the roads and steam-hauled trains.

During the early years of the 20th century a number of companies providing public transport amalgamated into a larger privately owned organization called the Underground Group. At this time they established some form of visual focus for itself through the use of a logo which incorporated a tram car in the centre, and underground railway lines leading to it.

The Group also developed a limited architectural vocabulary with its platforms and stations for the new tube lines, opened in 1906, all with the same station design using a consistent Underground lettering style outside every station. They were extremely publicity conscious and soon had produced free maps for handing out to the public.

In 1933 they amalgamated with all other underground railways, bus and train operators in London and formed one new monopoly public body, responsible to the local government authority. The company was called the London Passenger Transport Board, soon known to everyone as London Transport.

This transition from private enterprise to public service was marked by the gradual development of a trade mark which was finalized in 1933 and is still in use today. With the creation of a single public authority responsible for all bus, tram and underground railway operations in the capital, it was possible to develop a design ethic for the whole organization.

They were anxious to promote the public image of a progressive, efficient, caring and style-conscious company, at the same time as using design as a means of 'harnessing commercial methods to the achievement of large social objectives' – their words, not mine.

Good design could be good business, but for London Transport it also represented a major contribution by the company to the creation of a civilized and well-planned urban environment, and they used the best designers of the day to help achieve that objective.

In terms of product design their buses and underground trains were the most advanced and sophisticated in the world. Bus development culminated in the custom-designed Routemaster bus, and train development culminating in the fully automatic one that is running today. Other hardware specifically designed for them included ticket machines, notice boards, rubbish bins, light fittings, seats, signage, bus stops and shelters – just to name a few.

In terms of environmental design they created what was termed 'a new architectural idiom', a modern design form appropriate for central London stations, as well as one suitable for suburban stations.

Both designs set the familiar house style of London Transport for the years to come and were capable of considerable variation for different sites and structures.

Innovative designs were also produced for rebuilding ticket halls like Piccadilly Circus and the dramatic new headquarters for the organization at St James's Park. Even bus garages were now treated as part of their public identity and were used to give greater street presence to the company.

As far as information and graphic design was concerned they soon acquired an international reputation for patronage of modern graphic art by commissioning colourful pictorial posters to publicize the company's services.

Through the inter-war years the London underground station had become a popular showcase for avant-garde poster design. A particularly significant development in the company's publicity was the redesign of the geographical underground map in the early 30s into the familiar (easy-to-read) topological diagram; the concept of which has been copied the world over. Through its product, environmental and information design, London Transport was able to present a consolidated and unified message to the travelling public, that every care had been taken in providing them with the best possible service that was both easy and convenient to use.

This co-ordinated approach to design culminated in the development of the Victoria and Bakerloo Lines between 1969 and 1979 where each aspect of corporate identity was treated with the same sense of priority, and even earned the tangential compliment from Nicholas Pevsner that 'this was the most efficacious centre of visual education in England'.

These two new lines were designed as a package and reflected a new image of clinical efficiency appropriate to the world's first fully automated underground railway. And to top it all they even produced a design manual.

Since then, however, there has been a degeneration of design throughout the organization as a result of many complex factors.

Underground platform designers became preoccupied with superficial decoration, particularly in over-emphasizing the geographical sense of place of each location. Passenger information on both trains and buses became sloppy, unclear and unco-ordinated. Much of the hardware in use became treated as though it had nothing to contribute to the public perception of the company. And there was even gross disrespect for the one thing that the company had developed and established as representing all that was good in public transport, and in London – the symbol.

It would be over-simplifying the situation to say that all this happened as a result of one incident. However, during the 70s plans were made to move the control of London Transport away from local government to central government. Perhaps these political issues diverted everyone's attention away from their integrated design policy; it is difficult to be sure.

However, in the early 80s, with the passing of an act of parliament, a new authority was created – London Regional Transport. After that London Transport as a company ceased to exist.

This new body was charged with providing the most cost-effective passenger service within Greater London. Part of its task was to make the main business of underground and bus services profitable so they no longer required public financial support, and in 1983 major transport subsidiaries were created – one called London Underground Limited and one called London Buses Limited. The creation of these wholly-owned subsidiaries which are operationally independent has presented LRT with the problem of formulating its own corporate identity for the future.

Because of these changes we now have the ironic situation that, whereas design was once used to draw together the activities of a number of different companies at the turn of the century and unite them into an integrated transport service during the mid-century years, the company is now faced with using design to help it handle precisely the reverse situation. That is to allow the controlled fragmentation of the business on the one hand, and still to present a co-ordinated transport service to the public on the other.

Managing that design change needs dedicated effort. That is why I was employed, and why I report to the Chairman of London Regional Transport, and not the Managing Director of either Buses or Underground.

Naturally that presents its own problems. As the subsidiaries learn to identify and manage their own design work they have less inclination to be directed by the holding company, particularly at a time when they are working to be independent. However, as you will see, we have addressed that problem, as well as many others.

THE TASK IN HAND

Since I have been with the corporation, I have been concerned with initiatives in three basic areas. First with making managers aware of design issues; second with establishing a constitution for design that accommodates the special characteristics of the company; and third with managing specific design tasks.

Let me start with the design constitution. That is now in place in the form of a company standing order; there are six others on a wide range of corporate issues. Standing orders are of great importance in our business and together determine how the company is run. A senior management's awareness of design issues is very much reflected in the fact that we do have such a standing order. I would like to go on and explain this constitution to you, and then briefly describe some of the specific design tasks we are addressing.

There are three aspects to our design constitution:

- The first defines the objectives for a corporate design policy.
- The second defines the scope of design within the corporation.
- The third describes how design is to be managed.

Let me explain these a little more fully.

The first aspect of our design constitution deals with policy objectives. The main benefit from having to define a design policy is that it makes you think about what its principal objectives are. In our case we have focused on five:

- First to ensure that within appropriate financial constraints, the highest possible standard of design is established and maintained in all aspects of our operation.
- Second to co-ordinate design at a corporate level.
- Third to ensure that within this co-ordinating framework adequate scope is given to our subsidiaries in order for them to develop their own design ethos.
- Fourth to develop the use of design as a corporate resource.
- Fifth to establish a framework for design management within the organization that will enable these objectives to be met.

The second aspect of our design constitution deals with the pervasive nature of design. All our design work falls into one or more of the following areas.

- Products, designed largely by design engineers and industrial designers, which are used to provide the service (trains, buses, hardware, ticket machines, electronic equipment etc).

- Environments, designed mainly by architects, civil engineers and interior designers, through which the service is provided (stations, depots, shelters, workshops, office building etc).

- Information systems, designed mainly by graphic designers and typographers, through which the purposes of the organization are communicated (signs, posters, advertising, reports, timetables, promotional literature etc).

In practice, the scope of design concentrates on those aspects which have a direct impact on customers and staff (particularly, but not exclusively, those aspects which have a visual impact) and on wider considerations, such as the impact of design on cost effectiveness, maintenance, flexibility, safety, availability of resources and materials, and so on.

The third and final aspect of our constitution deals with the management of design, which in turn involves three distinct, yet closely related, activities.

- First, to establish working design policies for the organization
- Second, to implement those policies
- Third, to monitor the implementation process.

Let me give a brief explanation of each.

The first step in our design management procedure is to ensure that the right policies are firmly established, against which to work. For this to happen there must be commitment from the very top.

In our case establishing design policies for LRT is the principal task delegated by the Chairman and Chief Executive to the Design Director and the Design Policy Committee. This Committee assists LRT, and its subsidiaries, in formulating their own design guidelines, and provides an auditing role ensuring that those guidelines are met.

As a committee of the LRT Board, the Design Policy Committee meets formally three times a year, with the full authority of the board on all design matters. Membership consists of four members of the LRT Board, two company directors including myself, and five outsiders from the design profession.

The second step is to ensure that established policies are implemented at both a corporate and subsidiary company level. At a corporate level by the Design Director, and at a subsidiary level by the Managing Directors of the businesses.

At a corporate level, the Design Director is responsible for the following.

- First, providing a co-ordinating function between the design programmes of the subsidiaries, and those of the corporate holding company.
- Second, establishing and maintaining a framework for design management within the LRT organization.
- Third, providing an advisory role over the whole range of design activities of the corporation.
- Fourth, reporting on progress and change to the LRT Board.

In addition, the Design Director manages:

- The corporate identity programme.
- The design of those physical resources at the centre which it is not appropriate to delegate to the subsidiaries.

At subsidiary company level, each managing director is responsible for ensuring the proper implementation of LRT design policy, and that of the individual business. Additionally, they see that all design management processes and procedures are observed.

The third step in our design management structure is to ensure that implemented policies are properly monitored. In order to ensure that all design work undertaken within LRT conforms to the design policies, we have established two working groups. These are the Design Working Group and the Property Design Group, both of which the Design Director chairs. It is important to understand the executive function of these two groups.

The Design Working Group reviews and approves all design work undertaken by LRT and its subsidiaries. This is not an advisory function but an executive one. The Property Design Group approves all architectural design work clearly identified with LRT. This too is an executive group.

In order to aid the whole design management process, the Design Policy Committee has agreed the issue of a number of guidelines which can be used by line managers. These cover a number of topics, and the first five are already written. They are:

- the scope of design
- the design process
- the design brief
- inside/outside designers
- designer selection.

Many others are planned.

At the same time as establishing all this, it has been necessary to get some practical work done. So where do you start in a corporation as big as this one, with devolved responsibility a reality, and a tendency to approach design work on local rather than globalistic terms?

LRT first started with a number of design audits. As a result of these, specific design tasks were identified. This enabled co-ordinated design programmes to be established. Once they were started it became necessary to monitor progress.

SOME CURRENT WORK

You may be interested in some of the work that is actually going on at the moment. Some of it was started before I came to LRT, some has been started since my arrival, and some has been identified as being important but still remains to be addressed.

Even though the projects are very different from each other they all have a common purpose – to produce benefits through design excellence. The investment LRT is making in design is resulting in clear and tangible benefits in providing the best possible passenger service and working conditions for staff. Those benefits are assessed in three ways:

- public benefit
- staff benefit
- organizational benefit.

The projects range from quite narrowly-focused uses of design like the colour of the handrails on buses for the partially-sighted to the comprehensive design of the new Docklands Light Railway. They show the wide scope of design in LRT, and the enormity of what LRT does. In print, for example, there are more than 2000 new items produced every year with runs of up to 8 million copies. The investment in design-related projects over the next few years is likely to be in the region of at least one billion pounds.

Although the following projects have only been chosen as a snapshot of current activity, and not as examples of 'good' or 'good-looking' design (that is a judgement that must rest with others), they do demonstrate that the company is committed to re-establishing a standard of design excellence for which it was once so famous.

LRT's basic design philosophy is very similar to that of its predecessors: namely, to use the best contemporary architects and designers on those projects that contribute to improving the system for the future, while at the same time maintaining the best of its heritage. There is a positive conviction that the best of the new can live side by side with the best of the old. This is an uncompromising attitude that places contemporary excellence before pastiche and fake historicism.

Station design

Naturally, this attitude does not ignore the fact that London Underground in particular has a first-class architectural heritage. In fact it leads to two basic approaches when it comes to station design – restoration and remodelling.

Where appropriate, stations are restored to their original form, as was the case with Baker Street. On this occasion there was sufficient sound fabric remaining to make this approach both feasible and desirable.

When restoration is not possible, nor appropriate, then the station has to be remodelled; and in some cases that remodelling is based on the visual theme of the original. This was the approach that was adopted when the platforms at Piccadilly Circus were rebuilt a few years ago, the ceiling hoops and station naming of the early building determining the creative direction for the new one.

There are 273 stations on the Underground system, built over a period of 125 years, in different styles, by different designers, and currently in varying states of repair. Some of them are superb examples of railway architecture, designed by people of great stature, such as Leslie Green and Charles Holden. LRT has now created a 'Heritage' list of 45 stations that are of significant architectural merit, and, where it is financially practical, they will be conserved.

The evidence of London Underground's remodelling programme is there for all to see at many locations, particularly in central London. In many cases very well-known artists have been employed to work with the LRT architects to develop the creative direction for this programme. This initiative has created a resurgence of public interest in the Underground, and projects such as Eduardo Palozzi's Tottenham Court Road, and currently Robin Deny's Embankment have caused controversy and comment, which I would like to think will continue.

It is not always the famous that are employed for such work. Annabel Grey, for example, is an up-and-coming designer who has undertaken two schemes – at Finsbury Park and Marble Arch. Some schemes are more restrained in approach than others, for example at Paddington. The surface decoration for this station (designed by David Hamilton of RCA) is based upon Brunel's engineering drawings.

Rolling stock

High-quality design in train rolling stock has always been associated with the London Underground. For example, the 1923 standard stock with its motor and driving gear at the front car; or the 1938 rolling stock which represented a major leap forward with the traction unit mounted under the floor – a remarkable design achievement; and the 1973 definitive evolution of the earlier design, which has now become a great classic. Not only engineering and body design, but also interior design has been of great importance to the company.

It was therefore natural that when a new generation of rolling stock was needed for the 90s and onwards, the project design team should not only consist of electrical and mechanical engineers, but also an industrial design consultancy – David Carter Associates. Three new prototype trains have now been built for market research with the public, each with different heating and ventilation systems, as well as seating/standing ratio variations.

From a technical point of view, the train is very advanced in terms of power consumption efficiency, and body construction which uses the most advanced forms of aluminium extrusion techniques and phenolic moulding processes, while from a user's point of view the electronic passenger information systems, and indeed the interior design, generally, are also setting high standards.

Train livery

Vehicle livery is always an important way any company reinforces its identity with customers. This is no less true with transport operations generally, and with LRT's predecessors in particular.

The 1912 driving car has a livery which emphasizes the doorways and the 1938 livery has become a classic that many remember today, particularly as it has only just been removed from general service.

Continuing the tradition of creating a distinctive livery, while at the same time trying to come to terms with the modern menace of graffiti vandalism, was the task that was set designers Henrion, Ludlow and Schmidt. In addition, they had to develop a visual system that not only accommodated the fact that there were four basic types of vehicle on the system already, and that each had a different style of product architecture, but that not all of the trains could be repainted simultaneously.

The new scheme, which is still very much in its early development stages, uses the company colours on the front and doors, and the body colour can either be unfinished aluminium or a light grey paint.

Ticketing system

A new ticketing system is being installed in all London Underground stations. The features of the system include:

- modern, self-service machines which issue a wide range of tickets as well as give change;
- more spacious ticket halls for travellers and ticket offices for staff, who will also enjoy a more secure working environment;
- automatic ticket validation on entry and exit at busy central London stations;
- lower maintenance and operational costs;

* and a significant reduction in fraudulent travel.

Station signing

Historically, signing on the Underground has been world-renowned for its clarity of information, bold simplicity of layout, and quality of mechanical construction. Now a new signing strategy has been developed which builds on those traditions of the past, and presents very clear passenger information. A pilot installation is being tested at Victoria that deals with signing problems above and below ground, as well as those related to modal interchange. In addition to bringing a sense of order to passenger information, the benefits of those using the system will not only be in terms of faster assimilation of information, but also less congestion at critical points.

Posters

There is a similar story to tell about posters. Traditionally, London Transport used to commission the best graphic designers and illustrators of the day to produce posters that promoted events as well as destinations.

The standards set by the organization were very high, and the designers responded accordingly. Great names now of course, but not so well-known between the wars. Names like Sutherland and Eckersley, Nash and Moholy-Nagy.

This tradition of commissioning such posters has been recently reinstated. High-quality graphic talent is employed to produce posters which are both decorative and travel suggestive. Some names you may know, others you may not. The artists and designers include Kay Gallway and Alan Fletcher, Jenny Tuffs, Huntley Muir and Benouit Jacques.

Passenger security

In today's social climate, we must make special provision for passengers' security. The problem has become acute enough for the company to undertake a new major initiative. This has been made necessary first because of the unacceptable level of crime on the Underground, second to safeguard future revenue, and third to fulfil London Underground's obligations as a responsible public organization. Special waiting areas have been designed with focal points where passengers can get assistance, and alarm call points to be used in times of emergency.

Corporate identity

On a more general level, the objective of the recently started Corporate Identity programme was to develop a visual system that properly reflected the changing nature of the organization to all those with whom it comes into contact, particularly passengers, staff and opinion formers.

Potential benefits were identified in terms of a better understanding of how the company is organized, what its aims are, and clearer levels of communication. In addition, the programme helps to underline the corporation's position as one of the premier transport organizations in the world.

Bus design

As such, we have been concerned with many aspects of bus design. One recent major task has been to develop a bus entrance and exit that facilitates the introduction of one-person operation without compromising timetable reliability.

The new bus entrance has resulted in faster boarding times, shorter queuing time, better adherence to schedule – and enabled the introduction of one-person operation. The new entrance and exit will also give up to 1000 more bus service hours per day.

Publicity

Publicity material must work hard if it is to be easy for passengers to use, provide high levels of communication, and strengthen corporate marketing objectives. Market research has demonstrated that passengers find material easier to read and understand, more relevant for day-to-day travel requirements, as well as making the system easier to use. The organization has achieved better communication as well as major cost savings, conservatively estimated to be £1 million per annum.

Docklands Light Railway

In building the Docklands Light Railway the task was to increase land development in east London and provide visitors and residents with a cost-effective means of public transport. Passengers benefit from this crucial link between central London and the fast-growing Docklands area. While the organization has been able to introduce more efficient working practices, and achieve significant revenue generation.

Communicating the system

There are many projects which we have only just started to address, and there are some which are still waiting in the queue. For example, the complicated task of trying to communicate the intricacies of the bus network to the general public. To do this well would be a major step in getting more people to use the system.

Lighting

Considering that much of our railway system is underground, and has no natural light, our standard of artificial illumination leaves much to be desired.

Hardware

The design of a new range of hardware, for both passengers and operators, is almost complete and will be phased into use during the next two years.

Passenger information

We are already making significant strides into the area of passenger information, although much is still to be done, particularly with service change information, as well as safety and statutory notices.

Retail

Nearly 3 million people a day use the Underground. That is a very significant captive audience. Our estimation of the income from a new retailing initiative on the system is between £80 and £100 million. We are well into the design of a new retailing proposition for travellers.

CONCLUSION

Having said all this there are still many problems to resolve.
- Communication can be a problem (size of company, and its complexity).
- Tradition can be a problem (the company has many specialist groups that have a tradition of seventy years behind the way things are accomplished).
- Commitment or its lack can be a problem (a constitution is one thing, commitment and co-operation are another).

However, the company is beginning to resolve these problems, it has established a constitution for design to be influential at the highest and most pervasive level, it does have a framework for establishing policies and managing work, and different parts of the company are addressing complex design programmes.

It is early days in the renaissance of design at LRT, both at a management and practical level, but even so, some top executives are beginning to have expectations of design. And by having those expectations, they have taken the first important step in changing a vision into a reality and putting 'design into management'.

John Clothier is the
Managing Director of
C & J Clark Ltd which makes
and sells shoes worldwide
through a number of
operating companies
including Clarks, K Shoes,
Ravel and other chains in the
UK, Australia and France.
He is committed to the view
that design has a major role
to play in the successful sale
of men's, ladies' and

Designs on Your Feet

children's shoes.

I am not going to deliver a lecture about the importance of design in business; Terence Conran does it. Professor Peter Gorb of the London Business School gets paid to do it. I heard and believed what they said, so decided to see what could be done to increase the understanding of design in our business.

First let me tell you a little about the company of which I am Managing Director. C & J Clark Limited was founded in 1825 by my ancestors, Cyrus and James Clark, hence C & J, sons of a sheep farmer in Somerset.

Starting as a cottage craft industry 160 years ago, it developed into a mass production industry a hundred years ago when a factory was built on the site of the original farmhouse in 1890. The company continued as a single factory unit producing about a million pairs a year until World War II. Subsequently, the company expanded rapidly during the 50s throughout the West Country with the building of ten factories. Retail expansion, in the shape of the Peter Lord chain, took place in the 60s and most of the international growth was achieved in the 70s.

Today C & J Clark controls four main subsidiaries, namely:

- Clarks
- K Shoes
- C & J Clark Retail

A new subsidiary responsible for Ravel, Peter Lord, John Farmer and a chain of shops in France called France Arno and, overseas, the subsidiary that looks after our business in the USA where we operate the Bostonian and Hanover brands and in Australia where we manufacture and market brands such as Footrest, Clarks and Hush Puppies.

We employ 24,000 people, make 23 million pairs of shoes per annum and sell through 1600 shops. Our turnover in 1987 was £600 million.

I am going to tell you about the challenges of having designs on your feet and the way we have worked with the London Business School to open the aesthetic eyes of our organization. This work has been done within Clarks Shoes Ltd, the subsidiary responsible for the Clarks brand.

There are those of us who cannot design at all, there are some of us who can design in the mind's eye; and a few of us can express our designs on paper or as a model. Shoe designers, our shoe designers, design some lovely shoes but when the rest of us to go work to realize those creations, things easily go astray.

'Give me five minutes, and I will design you a better shoe than that.' How many times have I heard that said to me by salesmen, telephonists, people on trains, on holiday, and particularly by mothers at dinner parties, or on any occasion when they find out where I work. It is a bit like being a doctor, once they find out what you do, they know what is wrong with the National Health Service, and for us they are convinced that we don't care about their particular feet which are, of course, entirely different from everyone else's feet. 'Yes, you're quite right,' we say unclenching clenched teeth, forgetting for just that moment that the customer is always right. What they mean is 'I will tell you what I like on my foot' and that, of course, will vary with every individual person. They may very well be able to.draw a pleasing design, but will it fit, can it be made, will other people buy it and what will be the right price?

Let us consider what a shoe has to do:

- it has to be comfortable
- it has to fit
- it has to cope with innumerable surfaces
- it has to be flexible
- it has to be protective
- it has to look good
- and just as important it has to feel good on the foot and in the hand.

This is very much a 'hands on' industry, but before we start on hands, for a moment let us consider feet – however unpleasant that may be.

To the extreme thinkers they are the 'earth contact' – the point at which we are attached to the earth's forces and therefore, the plug from which we get our energies; to the reflexologists the foot is the gateway to our bodies and ultimately our minds. To most people, and probably to you, they smell, they sweat, they swell, they hurt and sometimes become completely unco-ordinated. Have you ever considered what an absolutely ridiculous object a foot is? Feet are not symmetrical, at least a foot is not. Together they might be. Feet, unlike hands, are considered unattractive to most people. Feet are more likely to develop something wrong with them in time, like callouses, bunions, ingrown toenails and, lastly, feet are much to small to hold up our weight.

Outside our headquarters in Street, Somerset, there is a Henry Moore work, *Sheep Piece*, and when Mr Moore came down to put it into place, as he always insisted that he did, he said that in his youth when he was at Art College, he had to build a torso, not full of holes as they are today, but one with weight distribution as per the human frame; this he did but in order to get the torso to stand up without falling over the feet had to be 34 inches long. Just imagine what we would look like with feet twice the current length.

So why are we able to stand up at all? It is because the foot contains a number of devices for balance, nervous systems, and motorization which enable us to walk. So a shoe has to cope with all these things: protection, flexibility, dynamics, comfort and be fashionable. That is asking a lot. Yet all these factors have to be taken into account when a shoe is being designed:

- its impracticable size

- its impossible shape
- its biological necessities.

As I said earlier this is a 'hands on' operation and I would like now to tell you what I mean by that. In this business you have to have a feel for the product and maybe that is because you are feeling the product most of the time.

A shoe's life starts with the customer. And by customer, I mean consumer, we are a consumer-orientated business, we have to be. Therefore, our first priority is to understand the 'need' of the man, woman or child often with the help of market research. For example, our children's team must get inside the minds of four-, six- and ten-year-olds and understand their fantasies. There is not just one type of girl or one type of boy. Segmentation starts young. The marketeer then from this information issues a brief to the designer for a particular customer requirement. Long gone are the days when the customer got only what we could make.

The designer starts from this design brief together with her inputs from the 'current' fashion scene and the predicted fashion influences that she will have picked up from her travels to the style centres of the world. Their task is very much to anticipate what will tempt the customer whose needs we are trying to satisfy. We work from six to twelve months ahead so our predictions must be right, striking the balance between flair on the one hand and the commercial restraints such as price of heel height imposed by the customer on the other. The designer will then work on several models to give options to fill the requirement: do the shoes meet the customer requirement? How can they be made by the factory, and by which factory, and what will be the price and does that meet the price requirement of the market place? Then prototypes will be made.

Shoemaking consists of three main parts: cutting where the patterns are cut like pastry on the flat; closing or stitching where the pieces that comprise the upper part of the shoe are stitched together and any design or patterning on the top of the shoe is machined on and finally the making where the shoe is turned from two dimensions into three or even four. The shoe is given height, shape and hopefully style. Just imagine how many hands are involved in the process so far, but it does not finish there. The shoe must be boxed, packed, stored, dispatched and sold. Here again hands are in use all the time. That is the process.

What are the design parameters? The customers' priorities are hotly debated and expensively researched by our marketing departments and market research agencies. Historically most women, but less men, put style as the only requirement. Others admit you must be able to walk in the shoe. We have adopted this as an important tenet of our business. We have a sense of failure if the shoes are kicked off under the table at dinner. Many shoes look wonderful but fall off the foot the moment you put them on. Some you cannot even put on at all. Women have sacrificed comfort for fashion and that has been so since the Middle Ages. Men have not been as fussy which is why on average they have much healthier feet than somen. As many as 60 per cent of women over sixty have foot complaints and only 20 per cent of men. However, as fundamental changes in our society take place so may attitudes to shoes. It is probably true that there is a shift taking place among women where looking good was an essential prerequisite to feeling good. Witness the New York female executive who walks to work in trainers and dons her court shoes on arrival at the office.

Having said all this, however, we must not get fixated by the foot. We must now move on to see how we try to blend fashion with function. This we do by building teams. Our ideal is a team of four:

- a designer
- a marketeer
- a production engineer
- a factory manager

Most of our designers come from art colleges and polytechnics, and are moved straight into the design department. Most of our marketeers are graduates from university who will have spent at least two years working in the production process through factories and in the retail side of our business. There is a weakness here; we expect range builders to be articulate and train them to be more articulate and aware of the market, while we neglect the development of these skills in our designers. We plan to make changes here in the future. Our production engineers are usually people who have progressed up through pattern cutting or design assistants or shoemaking. Many joined us as technical students. Our factory managers are mostly graduates who have progressed through the production process of line supervisor, production and administration superintendent, and have spent some time in marketing.

These teams are essential to the business and much of the training is to do with team building and team leadership. All must be able to comprehend what is meant by good design. Each person in the team needs an aesthetic eye as an appreciation of what makes good design. Everyone has different ideas. When it comes to fashion or rather what is fashionable, you are into the world of personal definition and extraordinary aberations.

Here are just three views. For Oscar Wilde 'fashion is usually a form of ugliness so intolerable that we have to change it every six months.' For Coco Chanel 'fashion is simply a matter of proportions.' And the one most applicable to us with which I frequently beat myself: 'English women's shoes look as if they had been made by someone who had often heard shoes described but had never seen any' (Margaret Halsey – an American authoress who wrote in 1937 a book entitled *With Malice towards Some* and certainly towards us).

In the beginning, I talked about the ones that went astray. However good an initial design, the 'hands' through which it passes can deliver it in its full glory to its intended customer or have subtly or unconsciously stripped it of its integrity. For us, the mindless machine that faithfully reproduces the original concept is not an option. Each design, each size, each colour, in fact, every pair passes through a myriad of hands; and behind those hands is a pair of eyes which may instinctively see the difference that each detail makes to the whole. All to often they may not see at all.

Designs that work are right in every detail. If we can raise the awareness of our managers and of our workforce that good design lies in the last 10 per cent of detail than we would make good shoes happen much more often. This is a monumental task but we had to start somewhere and we decided that we should choose those managers with the most influence on the business as a whole in this respect and so we came to the London Business School to help us. They came up with the

'Delivering Design Course', which enables managers to experience first-hand some key aspects of the product design and development process normally undertaken by professional designers. By 'first-hand' I mean 'by doing', because designing must be done to be understood.

Most managers operate either side of design development. This might be in the development of a market-related brief for design personnel to work to, or the translation of a design so that it can be manufactured effectively. Thus these managers are undeniably making decisions that have a significant effect upon the ultimate design of the product without much design knowledge. A course that provides practical knowledge of design development begins to peel away the layers of mystique that surround the designers' tasks. With increased understanding comes the opportunity for improved communications and the chance for managers to assess their contribution to the design process against better criteria.

Questions like 'Was the brief the right one?' can be discussed more positively between a range builder and a designer, when the range builder has experienced something of the designers' tasks. The designers' contribution to the discussion also improves when they experience the range builders' tasks: but that is another story. Similarly a production engineer who has had the experience of developing a prototype which he designed will have a better understanding of the problems of early prototype development. 'You have to make it to see it – and to solve it, you must make it again', said one product engineer.

Individuals who participate in the course come from marketing, range building, resourcing, engineering and production, there are no designers among them. All of them play a part in the product development process. However, in addition, individuals from retail operations participate, because shops are a vital part of the product cycle. Twenty managers participate in each course, working in teams of five. The composition of these teams is carefully considered to ensure a reasonable representation from all relevant functional areas and all three shoe divisions, men's, women's and children's. A typical team would consist of range builder, area manager (shops), factory manager, production engineer, and either an individual from resourcing or marketing.

So how does this course work for all these disparate people and what does it produce? The 'Delivering Design' course consists of three phases: first, a practical experience of design development which has a strong likelihood of achieving successful design; second, a practical experience of the development of a design through a prototype stage; third, a presentation to the board of Clarks Shoes, chaired by an outsider, to reinforce the company's commitment to design. The prototype product is presented, accompanied by a documented description of the process that each team followed. Teamwork is a key element. Participants are encouraged to contribute to the team effort across a broad front, and the experience of phase one has the effect of cementing the team, because of their need to adopt unfamiliar roles.

Before I go on to talk about the three phases, I should stress that the course was not about the total design development process at Clarks. For example, the course does not begin with any assessment of markets or market requirements. Above all the course was not planned to give participants the impression that they were on completion of the course expected to adopt a designer's role.

Phase one is the key to the entire course, because for phases two and three to be successful, the outcome of phase one must be a design proposal to which each team is committed. They all must have the confidence to believe that their design can become a good product and that it is worthwhile to proceed to phase two, the prototype stage. In addition, we were working with individuals who are experts in their fields and who would be able to see many problematic issues in their design proposals. Because of these two issues, a great deal of time was spent in planning for, and developing methods used in, phase one.

We had to decide what managers needed in terms of visual skills. In our opinion, making visual judgements; and discussion on the finer points of detail, are skills a manager could put to good use; whereas improved ability to draw would be less useful. Consequently, the design development in the course is achieved through comparison and structured discussion with a very small quantity of drawing.

Phase one takes place here at the London Business School. There is an introductory session which explains the working method, for the next day and a half. Because this method is very new to the participants, we found the best way to describe it was by example, so Angela Dumas, one of the tutors, went through the same process as the participants from the buying of the shoes to the development of a prototype. After this participants begin by meeting a fashion design student; each team, as one participant described it, is 'issued with a fashion student', the task is to make a trip to Covent Garden, visit a number of shops and buy some shoes for the student and take them out to lunch. The routes are preplanned so as to get in as much as possible, the participants find on this trip they visit shops they have not seen before and gain a valuable insight into the attitudes and opinions of future fashion designers.

After lunch the teams return to the Business School, they have with them the shoes the fashion student chose. The students are most certainly not Clarks customers and their choice of shoe is almost bound to be considered 'way out', and maybe this is one of the faults of our management system that our managers are not close enough to today's 'way out' which is so often a pointer to tomorrow's 'way in' for their customers. Our biggest single style of the 80s started life as the Princess Diana shoe in 1981: its mass market appeal was to the over 50s.

The participants understand that using these 'way out' shoes as a starting point, their task is to develop a different pair of shoes, which is not a copy or even a watering down of the originals, but something that is entirely different. As a benchmark we give the objective of developing a pair of shoes which Clarks might sell in a year or two. However, this is not an absolute constraint. It is interesting that as we have been running these courses now for nearly two years the shoes that were selected on the first course two years ago are no longer 'way out' and have become part of the 'norm' in the current fashion scene. Who would have predicted that today our ten-year-old customers in the madonna mode wear their underwear outside their clothing and are wearing Doc Marten derivatives which last time round ten years ago were 'bover boots', just as who would have foretold that men's cosmetics would reach 50 per cent of the sales of women's cosmetics, or if they had, would have probably been burnt as a heretic?

But back to the benchmark – any shoe which is a commercial

proposition would be acceptable. The advantage of using the fashion students' choice as a starting point is that it removes subjective and aesthetic considerations, which although important can obstruct the progress of the teams in the limited time available.

The next task is to analyse the shoes in terms of detail, to describe the overall impression that they create, and to build up a range of other things (i.e. not shoes) that share aspects or characteristics of both detail and impression. The teams work to the same method which involves the use of one hundred slides, in five different categories of products.

Discussion, to compare slides in individual categories, results in the choosing of five slides which, when seen alongside the shoe, provide a family of objects that share certain characteristics. This is documented with words and diagrams and then distilled in physical details and perceived impressions. The process which begins in the early afternoon runs through till late that night. It is interrupted, by just one lecture on design, which reinforces visual literacy and provides the opportunity for the teams to return to the work with a fresh eye.

Slowly during the morning of the next day the development of the design emerges from the matrix and the simple diagrams, along with discussions on likely technical problems, possible solutions and choice of materials. During lunch procedures for phases two and three are discussed. Each team makes arrangements for their prototype stage, this includes the fabrication of packaging and display materials. Once arrangements are finalized the teams leave for Street.

Phase two is a day spent in Street, developing the prototype designs, and is co-ordinated by Victor Jenkins, a member of our design development staff who is very knowledgeable in this field. He had accompanied the participants to London, where he was on hand to participate in discussions and to provide help with arrangements for phase two.

Phase two is the day the team has to test, experiment and argue among themselves before the crunch day of phase three. It is the middle ground and probably the most tenuous. Here the impatience, the ideas and the blocks created by the team are put under strain and as they are left very much to their own devices, it is where some of the major conflicts may occur. They are also, of course, fast approaching their deadlines, and the problem of being creative has to be reconciled with being practical. Firm deadlines are an essential part of the creative process in business. It is where the product engineer, if there is one in the group, and there usually is, proves his worth.

Phase three is also held at Street. It lasts for a day-and-a-half and ends with a presentation to the board of Clarks Shoes and is chaired by an outsider. We have been fortunate to have had Rodney Fitch, Jane Priestman and Raymond Turner as chairmen. The focus of phase three is to set in context all the work of phases one and two and also to reflect on the design process and to consider, in the light of the course, where it interacts with a manager's job.

There are two seminars during the first day but in keeping with the practical nature of the course most time is devoted to teamwork. Their final task is to prepare a presentation of their prototype shoe, the packaging and display materials; and the presentation also includes a statement about the process they followed. 'It is the journey as much as arrival that matters.' As a result of these presentations the shoes may well find themselves in the range offering of Clarks Shoes for the future. Certainly, the shoes from the first two seminars have found places in the ranges and have sold well, but that is not the object of the exercise. It is useful nevertheless to get something back from the investment of man-power of our executives from the course.

We have now run four courses, eighty of our managers have participated. The courses themselves have improved between the first and the fourth. We have got better at running what we believe are pioneering programmes. In order to assess the value of these, Peter Gorb and Angela Dumas, who ran the course for us, are in the process of completing a series of interviews with some of the participants and from this we shall have some of our essential feedback on the success of the 'Delivering Design' approach. The next feedback will come from our sales, but in the meantime I am convinced that our factories and those of our suppliers who when greatly pressured to respond demonstrate increasing ability to do so.

We have always tried to be a design-led company. Recently we have been developing better-focused identities here, in the USA and Australia, carefully matching the shops through which our shoes are sold to our target customers. K Shoes, Ravel, Bostonian in the USA, Footrest in Australia, and now the Clarks Shop will appear in every major centre in the land.

In reality, design is being delivered by everyone in our business, from the way our shops are designed and laid out, to the way our shoes are presented and the way our staff are dressed to the way the shoe is cut, stitched and assembled. No amount of thoroughness at the top can alone ensure that the consumer in Truro or Thurso, Oslo or Oporto received the object as intended in the way intended.

In a world where visual communication is in an overwhelming ascendancy, businesses and those who prepare us for the commercial world need to adapt their priorities. I do not believe that you can teach people to be creative but you can develop their visual awareness, so that they see and react to the gremlins that eat at the heart of good design. So at Clarks eighty managers have been through the challenge of the design process, getting the experience of doing, under their fingernails for once, so as to increase their awareness of just what it takes to deliver good design. They have not become designers nor should they. But they do know much more how to make good design happen.

Good design lies deep down in the product as well as on its surface. So, too, must delivering design lie deep within a business. Appoint a design director by all means but recognize that the eyes that make things happen are the eyes that deliver every day, every week. Clarks and the London Business School have broken new ground together, made mistakes and learnt from some of them. As long as we have designs on your feet, delivering design must be a part of our company's culture, a living culture revitalizing itself to keep pace with its customers.

*Geoffrey Maddrell is the Chief
Executive of Tootal Group
plc, a major international
textile company. His paper
describes how design
provides a leading edge for
all of the products his
company produces, and
what they are doing to
ensure that designers and
managers work together
effectively in making best use
of design.*

Design and Strategic Change

I shall be talking about

- the industry, particularly developments over the past ten years;
- where Tootal fits in the industry; this naturally leads to the section in my talk where I shall outline the strategy we developed two years ago;
- key issues involved in the orchestration of strategic change;
- design elements which are so critical in this process of strategic change.

Before I go any further, I should start by giving you a shanshot of what Tootal is, so that you can see the rest of the talk in context.

1. Tootal is an amalgamation of many textile interests which happen to come under the same umbrella called Tootal. This process of amalgamation went through its most active phase in the 60s when the Calico Printers' Association (CPA) got together with English Sewing Cotton.

- CPA itself was one of the five most powerful companies in the UK at the turn of the century and brought with it a dominant share in

UK fabric printing together with an array of overseas interests which
I shall show later are of major importance.

- ESC was itself an amalgamation of cotton spinning and sewing
thread companies, the bulk of profits coming from American
Thread. It had also diversified and moved downstream, acquiring
Tootal Limited with its brands and man-made fibres and Barlow &
Jones with its household textiles.

- Unfortunately substantial proceeds from the remarkably profitable
and original product, Terylene, were not used by CPA in further R&D
but in wasteful diversification.

- The perceived reasoning for this merging of interests was very much
the mood of the 60s in favour of full integration. This, as we all know,
proved fatal.

- So, as Tootal entered the 70s it could be seen as a microcosm of the
UK textile industry with strong overseas associations.

- At the time we employed over 50,000 and had a head office of over
2000.

2. So what are we today?
We are not vast by multinational standards, with sales over £500
million; but we are profitable, with pre-tax earnings of £40 million.
And the total complement has been reduced to 14,000; with a head
office around fifty people.

3. Our businesses are broken down as follows:
- three in industrial thread
- one in consumer thread
- one in worldwide fabric
- a very specific fabric operation – batiks
- a worldwide specialized materials business
- and then three UK businesses which source products, some sourcing internationally – in clothing, homewares and stationery.

4. You can see from this brief picture that we are not solely a textile
business; we have recently been diversifying into product areas
outside textiles where there is a compatibility in marketing, distribution and sourcing.

5. But above all we are an international business, largely because our

markets are global and costs of production differ between one country and another.

Over 60 per cent of Tootal's profits are derived outside the UK.

Now let us look at where the industry has come from, concentrating particularly over the past ten years.

- When you look at the textile industry, you look at it worldwide.
- The most significant feature has been the shift of capacity from West to East, really over the past fifty years or longer.
- This all came to a head with a dramatic shakeout in the last ten years.

Let me start with the UK.

- Throughout the 60s and 70s the industry had experienced the vicious cycle of low productivity leading to low profits, in turn leading to low investment.
- When British industry was hit by the high level of Sterling from 1979, the textile industry was one of the most severely affected. Imports flooded in and whole areas of textiles were eliminated.
- The real culprit was obviously management and the absence of any real understanding of what the customer wanted and any commitment to marketing. The approach had been entirely production based.
- Fortunately, out of the ashes, marketing emerged and gained status. And from the 80s design was recognized, with designers being taken on across the industry.

Now let us look at Continental Europe.

- The picture in France and Benelux was similar to the UK. Great names like Boussac St Frères went into bankruptcy and had to be given government support. New management outside the industry emerged. Benelux no longer had a significant textile industry, outside of carpets in Belgium which is supported by government.
- The Germans escaped the worst of the debacle. They retained a textile machinery industry supporting a textile industry with a typically Germanic proneness for quality and investment. This industry is now exposed as it remained integrated and is no longer able to compete at the commodity end.
- The Italians moved from strength to strength. Why? Well, you will

not be surprised to learn, because of their very strong commitment to marketing and design.

- Recently we have seen the emergence of countries like Portugal and Turkey.

And what about the USA?

- Because of the size of the domestic market the tide of change came later. Imports had been taking much of the growth with the impact really starting to be felt from 1984.

- The USA now represents the major market from many developing economies in the Far East representing in some instances over 70 per cent of their textile exports – prices in the USA are favourable and production runs for Eastern suppliers reasonably long.

- The industry in the USA has been reorganizing – with companies reforming to concentrate on garments, home furnishing etc. However, their approach has been almost entirely domestic and defensive. It surprises me that they have not used this large domestic market to attack on a global basis.

And now we come to the real beneficiaries of change – the Far East.

- We are all acutely aware of the labour cost advantages in the Far East. That is only part of the story. The other has to do with investment. All the statistics we follow confirm that the gap between investment in the East and investment in the West is widening not narrowing. And it is not just the Japanese and the Newly Industrialized Countries (NICs).

- Let us take our experience in China. In Guangzhou we now have the most modern and well-equipped thread yarn mill in the Tootal Group. Manning levels are equivalent to the best in the Group. We wanted the best technology to ensure fully competitive quality by world standards. But the Chinese are also looking for the best technology because they are looking to the long term. We believe China is rapidly becoming the new centre of the textile world.

- The Japanese, like the Germans, moved up-market as the yen appreciated, investing in technology. Their solid commitment to R&D will find new avenues. However, their textile companies lack international structures and they are increasingly working separately from the big trading houses.

- Now the NICs (Taiwan, Hong Kong, Korea and Singapore) are having to move up-market as their currencies gradually appreciate. So we cannot assume that only the lower end of the market will be served from the East.

What do we conclude from this brief historical development?

- First that, while there must be overall growth in world textiles as a growing population aspires to better clothing, household textiles etc, this growth will inevitably differ. Obviously the major growth will be in the developing world, especially the East.
- Second, the textile industry is a totally international business. Production in general and commodity production in particular will increasingly come from the East, with cost advantages changing between countries – the lowest factor costs are moving to countries like China, Sri Lanka, Indonesia. Our thread companies across the world pick up all the latest developments in cost.
- Third, the major markets with the greatest buying power are still in the West.

Now let us look briefly at where Tootal has come from.

- We have been one of the prime examples of the shift of capacity from West to East. CPA alone was one of the largest UK companies at the beginning of the century, selling vast amounts of fabric worldwide. Most of that capacity has gone. Over the past decade we have been working with the developing world to establish quality production for worldwide distribution. Most of our yarn for growing markets and virtually all our grey cloth comes from the East.
- Tootal was dramatically involved in the final phase of capacity reduction in the UK and the USA in the period 1980–85, closing the major proportion of spinning and weaving.
- The diversification undertaken in the 60s and 70s into retailing failed completely, largely due to management and lack of commitment. There was a dangerous supposition that acquisitions could be left to their own devices.

However, we – the next generation – have been left an interesting legacy of associations, structures, knowledge and skills throughout the world.

- The most valuable is our international network which in some instances has taken years to develop. Let us look at the international spread of the Group.
- We have also developed a multiplicity of products and have product and service skills developed over many years to apply in a variety of markets.
- We have surprisingly good market positions, even by world standards – and we are determined to build upon them. In thread we are market leaders in the UK and Scandinavia – in the USA, Canada and Costa Rica – in the Philippines, Indonesia, Sri Lanka, Singapore, Malaysia, Australia, New Zealand, South Africa, Nigeria. In fabric we are leaders in shirting fabrics in Europe sourced from the East, in Australasia sourced out of the East, in UK vat-dye printing. In clothing, in back-to-school wear, in ties and in certain branches of Marks & Spencer; in homewares in certain niches; in stationery in direct service to large accounts; and in certain high-value specialized materials throughout the world.
- Finally, we have an invaluable low-cost sourcing structure. This is not just a matter of purchasing skills. It involves controlling the quality of production either through equity involvement (as in China, Korea, Indonesia) or management involvement (as in India, Sri Lanka, the Philippines) or simply in long associations (as in knitwear from Taiwan).

Now let us turn to the strategy we are pursuing which evolved two years ago after a thorough examination of the Group.

- Briefly, we believe we have a unique opportunity to link East and West. We have very little upstream capacity to protect. Years of association has meant Tootal is known and trusted in China, India etc.
- We can source at low cost, with full flexibility in investment.
- We can add value at stages in the value chain where service and control are critical (dyeing, fast finishing).
- We can develop or buy in appropriate technology.

- But above all we must control our markets and our distribution by having a thorough understanding of customers and by design, marketing, service and systems. This is where we must concentrate. Our Chinese partners will invest in the thread yarn knowing we can distribute it worldwide in good times and bad times.
- Let us look more specifically at our mission statement.
- This process of building a strategy is very much a long-term business, particularly when you are talking about an international business. It needs commitment, resolve, realism and a bit of luck. But it also needs time.

Now I would like to turn the talk over to the whole process of orchestrating strategic change with particular focus on design.

STRATEGIC CHANGE

In a few short words the whole process must start with commitment. We involved our full management in developing our strategy and have full commitment to our new mission. Then it is a matter of transferring that corporate direction into individual business plans which was the meat of our work last year. Of course you have to have a structure which meets your new growth objectives. But the most difficult process of all is the change in culture. This has been going on for some time. The senior management of the Group meets regularly to take stock – to ensure we have a common understanding and clear any blockages.

Our culture is based on: developing an international perspective. This involves a much higher level of sensitivity to different practices, attitudes and much more cross-posting. We must also become truly marketing-led. We are not sufficiently design sensitive. We must practise 'service' at all levels of the organization (even between one another). We must obviously demand excellence. Basically we should be stretching people towards their limits, which is sadly an impossibility. We are looking for an enterprise culture where initiatives can be taken throughout the organization; this means from my experience limiting layers to one – and only one – between the centre and the operating unit in the market.

- I hope we have now resolved the management issues involved in the strategic process. We at the centre are largely concerned with the development of a Group strategy. It is up to the units to convert this into action plans.

- As the great communicators know, brilliant strategies are worthless unless they are fully communicated and owned by all stakeholders. I shall be discussing corporate identity and communication later. Suffice it to say at this stage that we in the Tootal Group fully recognize as part of our process of change the need to come up front and communicate our strategy not only internally but equally importantly externally. And corporate identity is very much a part of that process.

- Finally, in any process of change there are timing issues. Strategic development is always a long-term issue, particularly if you are changing values in the process. However, we recognize that the shareholders are looking in specific financial terms and require evidence of performance in the short term. This is always a difficult balance.

DESIGN

Now I would like to give my own perspective of how design relates to this process of change. I should tell you at the outset that with an oil and paper background, design has not played a major part in my career development. However, marketing has. I am sure you will bear with me. In Tootal Group there is no doubt that design permeates everything we do. Design has received a great deal of coverage more recently and has increasing support. I would like to go beyond motherhood statements.

As I hope is now abundantly clear, our future lies in beating our competition in worldwide markets. To use the jargon of the day – it is in the markets that we must win competitive advantage, by understanding our customers and their needs and by developing appropriate strategies. We do not intend to own all of our upstream capacity which is capital-intensive and vulnerable to change. We intend to use our skills in sourcing at the lowest cost. So let us see where design actually plays a major part.

- In products and services we are constantly seeking to add value and thereby win competitive advantage. At the lower end of the product spectrum we must anticipate that Far Eastern competitors will increasingly take share. As I hope is now clear, these competitors are moving up market all the time so we have to be looking at the best in terms of design, quality and service to differentiate ourselves.
- This means that design must become imbued in our management approaches and attitudes. It has traditionally been seen as a backroom activity and we are now trying to bring it right forward. Indeed, it must be a key component in our strategic thinking rather than just a potential advantage.
- In our internal communications we are making a point of incorporating design. This is not simply a question of better communications but of a changing culture. We are also looking hard at all our facilities in a design context. We have recently taken a number of smaller branded clothing businesses out from under a broad clothing umbrella, establishing them in their own new facilities with design and corporate identity being very much up front.
- As I said earlier, this communication process is as important externally as internally. We are talking about changing an image which is a comprehensive process incorporating every visual point of contact. Let me give you an example of some of our corporate visual ID.

Design in our business

One of the tenets of our new culture is a commitment to the upper end of the quality spectrum in both products and services. Let us look at examples of design in our product range.

- Clothing: this is a specific area where we must anticipate increasing imports (there was a particular surge in the first quarter of 1988). We must add value and design and, other than supplies to Marks & Spencer, the major part of our product range is sourced from overseas.

- In homewares virtually all of our fabric is sourced from overseas. We are essentially converters. Once again a major part of the value added lies in design.
- Batiks: this has been a highly profitable business over the past two years, largely because of the increasing commitment we have made to design. We are supplying a variety of tribes in West Africa all of whom have different requirements in design and construction. It is the ability to understand these requirements and to convert customer need into an acceptable product of value that differentiates us from local production. It is interesting that local production tends to copy our designs. We need to be constantly ahead.
- In voiles we had a major share of the Middle-Eastern market some years ago, using Swiss voiles. These became uncompetitive and the Japanese took the market. We have developed a new range with an Indian supplier – and this has taken us four to five years. We will only succeed in displacing the Japanese if we have the right designs and a full range of products.
- In Hong Kong the major part of the business is sourcing fabric from China for sale into Australasia and the UK; we have a design studio which is also involved in the finished garment. Our manager there is convinced that you lead the way in the finished article on the back of which you gain fabric distribution.
- Equally, in the UK in shirting fabrics we are the most advanced in computer-aided design, linking Marks & Spencer with overseas fabric manufacturers.
- In Da Gama in South Africa where we are the most successful and the most profitable textile company we gain price primier based entirely on our quality, design and service.
- And in Calprina we are the market leader in vat-dyed printing of furnishing fabrics at the very top of the market.
- In retail, branded thread and haberdashery design is an important feature, not just in the product itself but also in the way it is packaged and presented.
- And in industrial thread supplied to makers of end products, 90 per cent of the design activity lies in the performance of the thread itself.

Price is not the key factor for our customers; their most important need is to ensure that their machines do not stop. Therefore, they are prepared to pay for the right performance. The other key requirement is rapid service. Packaging and branding also play an important part.

THEREFORE DESIGN

As I hope is evident to all from this brief review of our product range, design is a key element. As such we are in the process of elevating it. It includes obviously the aesthetic, in the form of colour, texture, construction etc; but in our businesses, where we are frequently supplying to other people who make up products, it is also very pertinent to the practical issues of performance. A great deal of our product development in thread is geared specifically to performance. We have sewing advisors throughout the world who are constantly assessing the practical implications of performance with our customers. It is our vital means of adding value. It must be part of the strategic process and discussion at the board level and be incorporated with marketing and production.

Design audit

Some months ago we decided to find out how good we were in Tootal in terms of quality of designers and in terms of our approach to the design function.

We found that Tootal is generally well-served in terms of the quality and capability of designers for the functions that are set for them. However, although they are well educated for the job, they tend to be selected for backroom tasks rather than getting involved with customers and at the boardroom level. Secondly, we discovered that Tootal does not serve its designers well in terms of conditions, training, support, involvement and career progression.

The main problems are associated with management attitudes to design and its place in the decision-making and action. Equally there is a problem of the capacity of the design function to meet these more demanding strategic and senior management requirements. Obviously

the action programme which is now underway must address the issues
of perception and structure of management as a whole and the issues of
capability of our design function to meet the task. We have already
initiated training of the designers and have added a module into our
senior management training programme specifically aimed at ensuring
much higher levels of sensitivity to design by senior management.

Promoting design in Tootal

In addition, we have formed a policy-making committee consisting of
the heads of the businesses where design plays a major part, some senior
outside advisors with a major voice on issues of design (of whom Peter
Gorb is one) together with myself and the head of corporate affairs. The
purpose of this committee is to supervise and promote the best standards
and to ensure we have a change in attitude throughout the group. We
will have an audit function of not just the process of change but also of
the products themselves. Secondly, we have introduced a system of
communication across the group between designers and management
called Designline and Designfax. Clearly the objective is to ensure that
whatever happens in terms of design in one part of the group is
communicated throughout. Thirdly, we have recently opened a new
London showroom which is the front window for the Tootal Group. On
the one hand individual businesses use this showroom for their customers on a day-to-day basis. It is also very much a corporate showpiece
where we try to display our products in the front window and welcome
constructive criticism. And finally, as I mentioned earlier, we are putting
a great deal more emphasis on training of both designers and management.

Now let us look at design and the corporate message.

Design and the corporate message

Here our main objective is to change the perception both inside and
outside Tootal in line with our international and market-led objectives.
We are still seen as a domestic company by many audiences. We are still
identified, because of the association with a brand name, with shirts and
ties. Other than the message itself there is also the issue of where to

deliver that message in an international group. Tootal is perceived very differently but also very positively in the Far East and in other parts of the world. Finally there is the issue of the dichotomy between a number of our businesses which sell to other manufacturers and those that sell to retail.

Corporate visual identity

Seven years ago and before my time we had visual anarchy. A great deal of attention was paid at that time by our corporate communications supported by my predecessor to implement a new visual identity, worldwide. It was at that time that the new corporate 'T' came into being. And there is no doubt that we have made a great deal of progress since then.

However, we still have the issue of the Tootal name which is identified with certain brands and has a different meaning in the UK from the Far East, and Middle East and the USA. Notwithstanding the difficulties we feel it is better to change perceptions than to throw out what is historically a splendid name. Therefore, we have decided that Tootal will be used as the Group name and as the name to describe our various businesses like Tootal Thread, Tootal Fabric, Tootal Clothing. At the same time we will not use Tootal in brand names or for companies which are identified with brands such as Osman.

We have also decided to use to the maximum extent possible other visual tools such as on the side of trucks, buildings etc to reinforce the membership of Tootal of all our units, especially those that are in the non-textile areas. Therefore as we have a wide variety of products in a wide variety of countries our policy is to make it clear both internally and externally that units belong to a larger group which co-ordinates its strengths. We intend to endorse operations with membership of the group with a light hand and in a variety of ways.

What about the problems?

First of all we have the issue of how we handle this endorsement. This is currently being addressed. Second, we have the substantial issue of gaining full management recognition and commitment, not just at the top but throughout the organization, about the importance of design. As this is obvious at a strategic level it should not be an insurmountable problem. Part and parcel of this exercise must be the more effective use of our design resource. This is all linked up with training, both for senior management level and for the design function. Finally, we have the problem of ensuring that design is incorporated into our strategic thinking and our whole management process.

CONCLUSIONS

I hope if you had any doubts about the international nature of the textile business that these have now been banished. It is foolhardy to assume in any domestic environment that textiles is not subject to international competition. Fortunately a long and painful history has left Tootal reasonably well-positioned, without capacity to defend in uneconomic environments and with full flexibility to source worldwide. However, our success will depend on really controlling our marketing and establishing distribution in the major markets in the world. Production will increasingly come from the East. This means that our investment must go into adding value in the markets as well as into service and fast finishing and into constantly improving productivity in areas where we do have our own plant. Design is the key area of differentiation in our markets and it is the key to communicating our own identity.

Derek Lovelock is the
Managing Director of
Richards, the women's
fashion chain of retail shops
which is a subsidiary of the
Storehouse Group.

Design Management at Richards

In the summer of 1988 Richards were winners of the *Financial Times/ London Business School Design Management award. The importance of the award to Richards is that it was not just a reflection of our commitment to excellence in design, as part of the Storehouse Group with Terence Conran at its head. That can be taken as read. Nor was it a reflection of excellence in fashion garment design. That, to say the least, is a subjective judgement.

It was an acknowledgement of the way in which Richards manages its design strategy, and communicates it to all areas and facets of its business. That design strategy is 'to ensure that all parts of the company unite to produce a coherent identity which conforms to a high standard of design – is appropriate to its market, and evolves in time with market trends'. This paper describes how the strategy came to be evolved and how it was implemented. First some history.

Richard Shops was formed in 1936 by John Sofio who modelled the business on the US company Learner Brothers. The name came from Sofio's brother Richard. By 1941 the business was bankrupt and Charles

Clore picked it up for £45,000. In 1949 he was able to sell the business on to UDS for £800,000. From 7 October 1983 Richard Shops Holdings became an associate company of Habitat and Mothercare – the final twist to a tortuous drama which had begun on 4 January 1983 when Basishaw Investments, a consortium led by Gerald Ronson of Heron International, announced its intention of bidding for UDS.

Richard Shops had been a subsidiary of UDS since 1949 and had often been referred to as 'The Jewel in the Crown'. It was for many years generally considered to be the leading women's wear fashion chain, with a wide middle-of-the-road appeal to the twenty to forty age range. Its advertising slogan 'such clever clothes' reinforced its clear image. However, Richard Shops reached its peak in the late 60s, and by 1972 when it was trading from 132 shops and earning profits of £2 million, the decline had already begun.

From the moment of the Habitat/Mothercare takeover it became apparent, as it often does in these cases, that it was not merely cosmetic changes that were necessary. What they had was a business where most of the stores were in a total state of disrepair, where there were no tills, and cash was held in wooden drawers – and kimball tags in ice cube trays, and where, for example, one of the buyers had never spoken to a colleague who had been with the company for two years. It was a business where each product range was bought in isolation and usually for a different segment of the market, and where sales staff had achieved justifiable notoriety for their ability to pounce on any customer who happened to wander near the entrance of a store. Hence the decision was made to start completely from scratch, and the Conran Design Group was drafted in to work with the new Richards board in mapping the blueprint for a chain of 'uncluttered stores' offering stylish clothes for the twenty-five to forty-five year old woman.

From the very beginning, the underlying principle was Terence Conran's dictum 'one pair of eyes' – by this he meant that the varying talents and creative inputs involved in designing a new store concept should combine in a way which would look as if one person, and one person alone, had done everything; with the target customer clearly in mind.

The strategy for the redesign of Richards covered the following categories.

1. Corporate image – the external appearance of the stores, brochures, advertising, publicity communications, press shows and all sales promotions and literature.
2. Interior design – from the general ambience and layout of the stores, to the meticulous detailing of the shop fittings, lighting, fitting rooms, interior accessories and furniture.
3. Merchandise presentation – label and hanger design, packaging, ticketing and all point-of-sale material.
4. Merchandise – embracing fashion forecasting, orientation, through to individual product design and range selection.
5. Internal communications – training ideas, and all internal and external documentation and literature.

The first major problem was one of logistics – how to manage the total redesign of the stores, merchandise, merchandise presentation and all internal and external communications, while of course at the same time running the business on an ongoing basis. This had to be completed within one year to achieve profit and cash flow targets by the total revamping of 147 stores beginning in February 1985.

We began by assembling a computerized 'master calendar'. This listed in detail every component of the business and the offer, from hangers to vases, from neck labels to staff and books, and a timetable for their redesign. In all it contained about 3000 separate items. Not only were we able to ensure that nothing was overlooked, but that everything was addressed in a logical sequence. It identified requirements for cross-functional involvement and all lead items were considered and accommodated. In addition, by identifying the necessary individual inputs required, the person charged with making the final decision and the date, we were able to ensure that management time was utilized to best advantage. A co-ordinator was appointed to control and update the master calendar, to monitor the performance, identify where deadlines were not being met, and assess the inevitable 'knock on effect'. The co-ordinator attended the weekly progress meetings and held regular reviews with each member of the board, to ensure that control was maintained.

Having tested and modified the shop fixtures and fittings in a mock shop, the culmination of nine months' work came to fruition with a testing of a prototype store at Wood Green in September 1984. Wood Green was chosen not only for proximity to central London but also because it was the simplest and most typical size and shape of store.

Despite the best-laid plans of mice and men to ensure that the closing, refitting and re-opening of Wood Green went according to plan, problems did inevitably arise that we had not foreseen. Accordingly each management function appraised every aspect of the operation and the results were assessed. Modifications were made where necessary and all functional inputs were combined to form a countdown schedule manual which was used as the bible, to ensure the smooth implementation of what had to be achieved in the following year. This ranged from the sale of old and very old fixtures, fittings, models, treadle sewing machines and staff cookers on closure, to the precise colour and source of the ribbon and scissors for the tape cutting re-opening ceremony.

A building and refurbishment programme on this scale would not have been possible had we not been able to control the selection, ordering and distribution of all the store components and shopfittings, by use of a specially designed computer model. All shopfitting elements were designated as primary and secondary components. Because the system was designed as a kit of parts, by inputting only fifteen primary components taken from a prepared plan some further sixty secondary elements, amounting to almost 1000 items, were scheduled on an individual store basis.

The plan was to rebuild and revamp 147 stores in a year. They were divided into four main groups of tranches. Tranche one consisted of thirty-four stores which were principally Wood Green look-alikes – that is, no major building work or no difficult planning consents, together with a selection of stores with difficult configurations and characteristics, which would represent the problems to be addressed in the rest of the chain. Tranche two comprised sixty-nine stores in the higher turnover bands. Tranches three and four comprised forty-four stores with a combination of lower turnover and a higher degree of difficulty. Each tranche closed in a period of quiet trading for five weeks on average.

During this five-week period, the opportunity was taken to send staff on carefully planned retraining courses, to equip them to operate the totally new procedures, systems and policies. They then returned to their stores one week before re-opening to familiarize themselves with their new environment.

This, then, was the scale of the logistical exercise and it should be seen against the background of running three businesses with different systems throughout that year. On the one hand the new Richards, on the other the old Richards shops, and third those shops which were part of the inevitable closure plan.

The same problem applied to our supplier base. If we were to move from a company that, in effect, purchased from wholesalers, to one that could establish a credible quality own brand, we had to quickly devise and implement a supply strategy. What we sought to achieve was a blend between two different sorts of manufacturer. On the one hand, large efficient but relatively inflexible concerns to whom we could not hope to be a major customer, but who would be prepared to dedicate a manpower resource to service our specific needs. And on the other hand, smaller concerns to whom we would be important and who would give us the element of flexibility we required. At the same time within these two main groupings we needed a blend of companies with their own design strength and companies with a low overhead base to make up our own designs and our basic items at the lowest possible cost.

The disciplined approach that we sought in garment design had to be matched in the production process. To this end we established a quality control department. They worked with the design studio and the major suppliers to produce comprehensive manuals to ensure consistent standards of garment specification, make-up, presentation and delivery. In addition, procedures were instituted to ensure significant garment technology input at every stage of the design, selection, manufacture and delivery process.

But what then of the merchandise? This posed a significantly different problem; less one of logistics, more one of people and creative people at that. Our aim was to achieve a wide range of exclusive merchandise, covering all life-styles with a consistency of design, price and quality, appearing to our target customer as if it were chosen by one person.

At the time of the takeover, a fashion studio of forecasters, designers, pattern cutters and machinists was established at Conran Design Group, to co-ordinate the development of the new Richards range of merchandise. This would be viewed against the historical background of a team of very powerful buyers occasionally lovingly referred to as 'Prima Donnas', who purchased in the main from wholesalers in and around the fashion district of Great Portland Street. To say that the process of trying to impose or merge the views of designers, with a fresh and clear idea of the new Richards concept, on buyers with a firm idea of what historically they have sold, was a difficult and tempestuous one would be something of an understatement.

The opening of Wood Green proved to be a remarkable and important catalyst. A team of the best field staff merchandised the stock in a colour-related fashion to suit the modular fixturing system. The inevitable happened. The front half of the store looked wonderful and the rear half contained all the odd items that had little in common and looked like the result of at least twelve pairs of eyes. Nevertheless, Wood Green proved that we were on the right track. Essentially, it was selling the same range as the rest of the chain; perhaps slighly edited. But the sales rose by 75 per cent due to the combination of the new store design and the attention paid to visual presentation. What then could be achieved with the right merchandise?

In fact our buying strategy was not to be fully formulated until after Wood Green. Prior to this, it had largely been a case of upgrading the merchandise and introducing some form of control into the buying process. Formulating a buying strategy to suit a store environment is – in most cases – the wrong way around. But perhaps at that time, the store designers had a clearer idea of the new Richards concept than the buying team, who were still deeply immersed in the trading pattern of the old Richard Shops. Whatever the reason, Wood Green was something tangible to see, feel and understand. What then did it teach us?

First – that our aim must be to achieve a total harmony between the product and the environment.

Second – that customers appreciated colour-related ranges and that excellent visual presentation was critical. We had to buy to make this easier.

And third – the modular systems, or 'bays' as we called them, separated the store and the merchandise into distinct areas. This gave us the opportunity to present clear, concise and different statements to the customer.

To implement our new buying strategy, we had to quickly and successfully blend the diverse talents and personalities of designers, buyers, garment technologists and manufacturers. To this end, we developed a procedure which not only disciplines the buying process, but is also an excellent means of communicating our buying strategy to all concerned. A change in perception of the role of the fashion studio was fundamental to this exercise.

It had never been our intention that the fashion studio design all our garments, indeed this was not necessarily desirable. Having restructured our supplier base, we were now using many manufacturers who had invested heavily in their own design departments. This was a valuable resource that we clearly should not ignore; especially as somewhere along the line we were undoubtedly paying for it. The brief of the fashion studio was to develop a number of fashion themes for each season – usually there are about six – commensurate with the average number of 'bays' in a store and a method to communicate these clearly to our buyers and manufacturers.

Their response to this brief has now become our standard procedure. After initial discussion with the buyers, they hold an 'orientation' meeting illustrating styles, fabrics, colours and silhouettes for each fashion theme. In addition, they produce a number of garments which epitomise the look we are trying to achieve. Once agreed, these then become the core of the range, or the reference point around which we can build, with suitable items from our manufacturers. This same full 'orientation' is then given by the fashion studio and buying team to each of our manufacturers individually, and 'orientation boards' created in any design studio are used for this purpose.

Consistent reference to the orientation boards throughout the buying cycle ensures a disciplined yet creative approach to the complex task of range building. In addition, the boards are then used as a training aid for all store staff, and as a basis to convey the message through our point-of-sale material to our customers.

Fundamental to the success of our buying strategy and proven by the Wood Green experience, was that the discipline necessary to achieve consistency in the design, selection and manufacture of our merchandise should be replicated in the manner in which it is presented in the stores. Only this way could we ensure that the customer saw the range as it was originally conceived.

Our market research told us that, on average, our target customers would visit us every three weeks. We therefore decided to establish a completely new department – that of visual merchandising, to design and manage detailed changes of store layout every three weeks in potentially some 200 or more locations with minimum disruption. The method by which this is achieved is that, before each change, the visual merchandisers are given a 'mini-orientation' of what styles and stories are due to be delivered. All aspects of current trading are taken into consideration to ensure that the best compromise between commercial need and visual appearance can be achieved. A model store is then remerchandised, and once this is signed off by both the store operations and buying department, the exact position of every garment is listed and photographed. A commentary is written which explains the logic and reasoning of the new layout, and guidelines are produced which cover variations in store size and shape. These are then compiled into a layout manual which is sent to every store, thus enabling the complete chain to be remerchandised, on the given day, to a consistently high standard. This is the process which is repeated every three weeks.

The restyling of the stores and the merchandise was accompanied by the redesign of our head office. Everything from the carpets and the colour of the walls, to the order forms and quality control garment seals, had to be in keeping with the corporate identity. Just as our stores and

merchandise are designed and presented to give the customer confidence that we have considered every detail, so the same concept in our
head office reminds everyone that everything must be done in the
Richards style.

A committee of four Richards directors meets monthly with Sir
Terence Conran and representatives from all disciplines in the Conran
Design Group, to discuss all new developments in design, marketing
and communications. In effect, this body is the custodian of the corporate identity.

The key question now has to be – 'What has been achieved?' The
results speak for themselves. Sales growth of 60 per cent in the first year
after the redesign. And sales growth of 24 per cent in the second year
accompanied by a 90 per cent rise in profits. We have seen equally
dramatic changes in customer perception. Recently we undertook considerable research into shopper and potential shopper reactions to key
aspects of the Richards offering.

The research (an extract from which is shown in the appendix)
substantiates our original strategy based upon 'one pair of eyes'; only
now we see it as achieving a measure of 'harmony'; harmony between
product, environment and service. It is a strategy successfully employed
I believe by other retailers like Penhaligons, the perfume house who,
interestingly enough, are prior recipients of this award; and by Body
Shop and our own menswear operation Blazer.

Let us once again revisit the original strategy:
- Richards is a relatively new company. It has only existed in its
  present form since 1985.
- In the two years following its acquisition in October 1983, every
  aspect of the business was reconstructed in order to implement, or
  support, the new business strategy.
- In the last three years, there have been a number of major developments, but these have all served to refine or enhance our approach.
  In essence, there has been no major deviation from the original
  strategy.
- This means that throughout the organization there is a clear focus

on our target customer, and a remarkable level of consistency in our approach to her.

- The merchandise strategy is designed to make Richards a dominant force in the chosen sector of the women's wear market.
- The merchandise range will be bought:
  - as though with 'one pair of eyes'
  - in good taste
  - to be fashionable without being high fashion
  - to be value for money
  - to be merchandised within a pleasant environment.
- The merchandise range will remain commercial, while aspirational to the market in general.
- The product and the environment will be balanced in such a way as to complement each other.

I believe that the main problem to face us in the future stems from our greatest strength in the past. We have achieved much by developing a strong, common vision and subsequently the means to manage and communicate it. We must now continually question, analyse and assess the results, and to react – adapt and change where necessary. For example, only now that we have achieved uniformity in the way that we develop and present our merchandise and manage our stores, can we accurately identify differences in store performance, due solely to specific local characteristics. Our task is now to make room for flexibility among all the discipline and consistency of our approach. We must find a way more accurately to tailor our offer to the requirements of particular stores, or groups of stores, while maintaining an overall consistency in appearance.

In a retail sector dominated so much by one player, and a big one at that (I mean of course Marks & Spencer), success through differentiation for the other competitors is difficult to achieve. We must accept that we are unable to obtain a sustainable competitive position in any single aspect of our offering. Not solely in product nor in environment or in service. However, by aligning the whole organization towards servicing a clearly defined customer we have, over time, achieved a total offering which is first, difficult to imitate and secondly valued by the customer.

This approach is the same as our historical stance, differing from 'one pair of eyes' only in the degree to which it can be made explicit.

As a business we have used design management to develop a discipline exercised within the company whereby personnel, at all levels, are able to gain a thorough understanding of the role of design within the overall strategy, so that they may have a consistent regard for this in their work. I believe this has given us an advantage in a very competitive environment, to ensure our continued success through differentiation, by satisfying our customers' needs and aspirations and by achieving harmony between the product, environment and service.

Sir Brian Corby has been the Chairman of the Prudential since 1982 and a director since 1981. He has spent his working life with the company. He has been Vice-President of the Institute of Actuaries and Chairman of the Association of Insurers. He has a number of external directorships including the Bank of England. He is currently the Vice-President of the CBI.

Implementing Corporate Strategy

Until three years or so ago my intellectual interest in design was no greater than that of most other people. In the shops I went into and at home I was surrounded by the work of designers. Some of it good – which in my terms meant that I liked it – some bad and some indifferent. I knew what I liked – when I saw it. Like everyone else I was under the influence of designers perhaps unwittingly, but I had not yet bothered to form a reasoned set of principles as a basis of support for the preferences.

I am now a more enlightened man – by no means an expert, but definitely more enlightened. The change started in the middle of 1985, when my awareness was heightened by a problem. Being a chief executive, problems are par for the course, but those of us in the financial services industry feel that we have been having more than our fair share recently. Some of you may be familiar with them, but in order for me to illustrate the whole picture, I need to sketch in the background first.

A number of things had come together in the late 70s and early 80s which caused radical and rapid change in the financial services sector of the economy. The banks moved into the traditional territory of building societies very successfully by providing mortgage finance, and had acquired a good share of that market. The building societies were then allowed, under new legislation, to provide loans for non-house purchase purposes, plus pensions and insurance services, and to offer money transmission services. Previously these were the activity of bankers and insurers. Customer attention was drawn to this by an enormous increase in advertising expenditure, particularly in the medium of television. The advertising expenditure on financial services has increased eightfold since 1980.

At the same time, dramatic changes were taking place in the world stock markets both organizationally and in the use of technology. Famous City names were losing their independence. Japanese, American and other foreign banks had moved into the UK retail financial services market in the belief that there was profitable business to be had. Deregulation was removing old and restrictive practices. New legislation to protect the customer was being introduced by the government. Further changes were in the making as retailers began to feel that they could sell financial services to their customers in a much more stylish manner than the banks, the building societies or the insurance companies. Also, if all this were not enough, then the customer – the likes of all of us in this room – was suddenly becoming more knowledgeable, more demanding of quality and service and more aware of his or her rights.

These events and issues were not unique to Prudential, they were common to all businesses in the financial sector in some form. Yet we knew that how we set about solving the problems and the skill with which we did so would distinguish us from our competitors and make the difference between surviving successfully or being relegated to a lower division.

In considering my response to these threats and opportunities, it was put to me that not only did I need to ensure the high quality of our existing skills in investment, accounting, marketing, selling and technology, but that I should also draw on the skills of designers.

Designers had, of course, been used for many years by the Prudential, for example, in print. But always at the tactical level. This was the first time design was being advocated as an important contributor to corporate strategy. So far, in the financial arena, this idea was relatively new.

Some building societies were trying to make their offices more user-friendly and so, in a small way, were the banks. Investors in Industry was perhaps the first important City business to embrace a proper design philosophy.

Manufacturing industry and retailers had, of course, seen the key importance of design for a long time. Philips, Braun, Sony, Olivetti, Next and Principles are ready examples, and in the engineering field, there are some spectacular examples of high technology in aircraft and bridge design. I deliberately mentioned several foreign companies there, because they have long used the power of design to sell their goods. For these industris the way the individual elements of an object were put together was clearly critical. Not just to make things visually attractive, but to make them simple and efficient in construction and use.

To me it seemed that the justification for the use of design skills at our strategic level must lie in the fact that they will make a clear contribution to improved economic performance of the business.

And before any of you cry 'typical business', have no illusions – no business can operate without making a profit. There are some people in business whose only conception of happiness is a positive cash flow, but that is no way to run a business except downwards. So in order to comprehend how big a decision this was going to be, let me tell you something about us.

Prudential Corporation is the third largest financial institution in the UK by market capitalization, with Natwest and Barclays both bigger. In global terms we rank fifty-sixth and just to show you where the biggest noise is coming from, the top seventeen are all Japanese.

Prudential Assurance was founded in 1848, so we have had 141 years to become Britain's best-known insurance company. In 1978 the company formed the Prudential Corporation as a holding company. All our services are sold through various subsidiaries, and this form of organization has given us far greater freedom of manoeuvre in the new financial services marketplace.

Prudential Corporation owns the largest UK life insurance company – The Prudential Assurance Company Limited; the largest UK reinsurance company, Mercantile and General Reinsurance Limited; and Prudential Property Services – a new large chain of estate agents in the UK. It is expected that by the end of 1989 it will have something approaching 1000 estate agency outlets in the UK. The services provided by Prudential Property Services cover not only residential property for sale but also commercial property, the purchase of land for development, and advice on mortgage finance and financial products provided by the group.

Prudential Portfolio Managers is responsible for investing the UK revenues of the corporation and is the largest investor on the International Stock Exchange; Prudential Holborn is a new division selling life, pensions and investment services to the high-net-worth individual.

Prudential sells insurance and savings products to individuals in the UK and in some thirty overseas countries (including the USA). Prudential Corporate Pensions provides companies, local authorities and affinity groups with a full range of pension services. Nearly half Prudential's revenues now come from overseas.

As you can appreciate, any decisions about the use of design in such a diversified company cannot be taken lightly, particularly since the corporation was developing a decentralized management style. Each business was gaining a degree of independence. How far should this independence go? Should each business area create an image in the marketplace which was uniquely its own, or should there be a single unifying identity which emphasized the name Prudential?

In other words should we be like Unilever, where its activities are known to the public through names like Birds Eye Walls, Batchelors Foods and Brooke Bond Oxo, or should we be like ICI which emphasizes the parent name in all its business activities such as ICI Fibres, ICI Fertilizer, and ICI Petrochemicals? To aid the debate Wolff Olins was asked in 1985 to take a look at us and to propose how we might best project ourselves in the financial services marketplace of the future.

So what images were we projecting in those days? The man from the

'Pru' was one visual association that Britons have always made with the name Prudential. The company's Victorian origins, grounded in the Industrial Revolution, have provided the Prudential with strong cultural associations. The very name, Prudential, derives from the quality and personality of Prudence, held in great esteem by the Victorians as a principle of wise conduct. The Prudential's main office is distinctly Victorian Gothic, with an imposing façade that suggests solidity and bureaucracy, not unlike government.

The familiar face of the man from the 'Pru' represented a huge, personal sales force on the one hand, and the impression of Victorian solidity on the other. These visual associations have been a strong part of Britain's commercial culture. However, until 1985, we had underestimated the power of a unified image which correctly portrayed the underlying spirit of the company. Our name was strong, but customers, and even many of our own staff, did not fully understand what it stood for in the mid-80s.

This was the first time that a study had been made to look into how we communicated ourselves both externally and internally. We took a look at all the identities of companies within the corporation. We listed all the different names of our companies and their product brand names, and looked at how they related or, more to the point, did not relate to each other.

Let me illustrate that point by reviewing some of those identities. Identities represented by logos generally fall into six categories. There are heraldic devices, such as those used by Porsche, Barclays, Philip Morris and The Prudential Assurance Company. There are those with old style lettering. Then there are abbreviations, key words, geometric shapes, and architectural symbols.

When all a company's activities are involved in one industry, in our case financial services, having such an array of symbols was obviously confusing. And the customer makes no links of association. The real irony of course is that the Prudential name is strong but did not appear in many of those identities.

The next step was to conduct some good market research to find out what people thought the name Prudential stood for. The interviews showed that most people knew our name, but were not always sure what our range of products and services was. Over and over, people said that we were 'traditional' or 'staid' or 'old-fashioned', a reflection of the Victorian past. Some people thought we were first and foremost an insurance company, but there was some confusion about whether or not we were involved in property.

In an increasingly competitive marketplace, such a confusion of impressions was not going to do our business any good at all. We immediately established that we needed a coherent corporate identity which would bring together all the companies in the group under one common banner. It would have to be flexible enough to be used throughout the company. This would have obvious benefit when poten-tial customers look to a name for an endorsement of quality of service before they buy the product. But flexibility would mean using the corporate identity consistently throughout the group worldwide. Above all, what we needed was a distinctive identity, which would set us apart from our competitors. But it was no use going ahead with an identity change until we had fully analysed our existing image, and this again came from our research.

We asked about the Prudential's character. Most commonly, people replied that the Prudential was honest, responsible, old-fashioned and staid. Honest and responsible are good character references, and worth retaining. But staid and old-fashioned? We needed to change those characteristics into secure, competitive and innovative ones – charac-teristics which we felt we already had, but were not appreciated.

We asked about the Prudential's personality and got responses like approachable, big and lumbering. Approachable is a desirable attribute but 'lumbering'? In reality we felt we were dynamic, but clearly we were not projecting it. For example, we have very advanced computer systems and we were already one of the top ten unit trust groups, despite being in the business less than two years. Also, there is nothing inherently good about being big. We needed to convey that we are a strong company; and when asked about our identity, the answer was a stony silence. As there was no argument for modification, we had to start from scratch and establish a completely new identity.

Now in case you are wondering what difference does identity make, I should explain that in financial services we do not differentiate our products on price and benefit as so many other industries do. A loan is much the same wherever you get it from. The great distinguisher for us is service. And within the company there was a growing concern that our staff did not recognize us as a major force in the industry. If you are not convincing your own staff, what chances have you got of convincing your customers?

The only place we could take heart from was the City. Financial journalists and business analysts all agreed that Prudential was at the leading edge of the financial services revolution. In fact, we were identified as a company that had the right management and resources to become a major winner against the competition. It was very warming at the time, but our customers were those we had to convince.

The design solution for our communications objectives combined the two key strengths – our name, and the embodiment of our name, Prudence. Prudence has been part of the company's symbolism since 1848, appearing at the top of the coat of arms of the Prudential Assurance Company, and in many other forms over the years.

In fact, there was a noticeable degree of unease among some of my colleagues when it was proposed to transport Prudence from Victoriana straight into the late 80s. Many ideas were discussed and at great length. Prudence was not always the front-runner. But she seemed right, as long as we could make her up-to-date, and striking.

As one of the other cardinal virtues alongside Justice, Fortitude and Temperance, it is Prudence which still continues to embody the very principles which the Prudential stands for. These are represented by the symbols which have throughout history been inseparable from her. A mirror . . . not representing vanity, but the ability of the wise man to see himself as he really is. A snake, the ancient symbol of wisdom. And an arrow for directness. With the minimum of graphic design, we took all those traditional symbols and brought the lady up to date. In a world that many feel is impersonal, Prudence gives our financial services a human face and a gentler touch. It is difficult imagining life without her now.

A confident launch of the new identity was our next task; but as I am perfectly aware that confidence is that warm, assured feeling you have just before you fall flat on your face, I was not going to take any chances – we had to make sure that launching the identity was handled well.

To us our audience was as much internal as external. We decided on a high-profile launch. We needed to communicate how important the company regarded the new identity, and to explain why. We needed also to re-emphasize how the marketplace for financial services was changing, and the threats and opportunities facing the company. And to impress on staff at all levels the importance of service to the customer. It was an expensive operation. It was big, grand and it created a great stir. No one was left in any doubt that the company meant business.

The launch took place in September 1986 in a week-long series of twelve presentations at London's Royalty Theatre to 8000 out of our total of 25,000 UK staff. They were chosen from all grades and all levels of experience from all over Britain. We used lasers, smoke, specially composed music, the whole razzmatazz, but only as a means of presenting the message in a dramatic way. We encouraged our staff to be frank about the new identity, and their reactions were highly positive and enthusiastic. As it turned out, this was just the sort of change that most of them had been waiting for. They were pleased that we had taken a traditional symbol of the Prudential and updated it. They liked the colour, the style, and many liked the fact that it was radical enough to cause a stir among their customers.

After each presentation we distributed a new corporate brochure and identity booklet. Within a week, over 400 local separate presentations were made in our offices around the UK and abroad, to the staff who were not able to visit London. These were carried out by 400 managers who had been briefed for the purpose and had already attended the London shows. They were given videos, flipcharts and even speaker's notes to ensure consistency in the message. Nothing was left to chance.

Nor was the follow-up. We needed to maintain the visual momentum: on the following Monday we introduced new letterheads, business cards and envelopes throughout the UK. After the staff presentations we made a similar one to the Press and other City opinion formers. As a result the press coverage was very extensive and, I am pleased to say, very positive.

So we had ourselves a new identity. Now what were we going to do with it? There is more to a corporate identity programme than changing old logos with a new one. Other opportunities present themselves, from advertising to merchandising and the environments in which our staff work and where the public meets us. Now that the identity was established, we set about using it in every conceivable way.

I now want to point out two other areas where design plays an important part with Prudential. The products in our industry are expressed in pieces of paper. Promises to pay on the happening of some event or at the end of a period of time. The systems to produce and service the paper are complex. We are as much concerned as any manufacturing firm (probably more so) that the design of these systems is sound and leads to economy in production.

Our second interest has been with architecture, although standing here today I am not going to enter any debate on the merits of Prince Charles' view on the subject – I would not dare. Our present head office in Holborn was built over the period 1874–1905 with additions in 1919 and 1930. The first building on the site took place at the junction of Holborn and Brook Street. As land adjoining it became available the office accommodation was extended. The development of the whole site has therefore been piecemeal. Its appearance as a coherent design is due to the great skill of Alfred Waterhouse the architect who was commissioned over the first thirty years to carry out the work. The 1919 addition was by his son, Paul Waterhouse. In the early days the building was advanced for its time in the standard of accommodation provided. The technology in use was also advanced.

Waterhouse designed many buildings up and down the country. For example, university buildings at Oxford, Cambridge and London, Manchester Town Hall and the Natural History Museum. Waterhouse was also the architect of twenty-two of the Prudential's offices in England

and Scotland. The Waterhouse architectural style was a form of corporate identity.

The accommodation is now becoming unsuitable for a company which is strongly based on the use of modern technology. So a four-year programme to bring the building up to the standards of the 21st century is just starting. However, we are fully aware of our heritage and have consulted the Victorian Society and many other bodies about all our plans. The Holborn frontage will be preserved and so will other parts of the original Waterhouse structure which are regarded as fine examples of his work.

I am sure you have not forgotten that I have already said 'for design to be taken seriously by management, it must make a clear contribution to the economic performance of our business'. Let me now rephrase this:

Good design can make a clear contribution to the economic performance of a business, and it must be managed with the same determination for excellence that is demanded of other skills such as marketing, finance and technology.

We accepted design in 1986 as one of the skill elements needed to develop our position as a leading financial services group. After two years of using design positively, primarily in its corporate identity form, it seemed only right to test that belief. Thus, in February 1989, we set out to see if design mattered to the man in the street and in particular if it mattered to our customers. The results of this second market research are intriguing.

We found that members of the public are very aware of the use of design in products, printed matter, and shop and office interiors. And becoming increasingly so. From the way design is used, they receive messages about the company. How they interpret these messages determines whether or not they wish to associate with that business. For example, some well-known shops are seen as dull and dreary while others are seen as bright and friendly. These perceptions are translated into the view of the quality of goods and services on offer.

They were equally frank about their impression of financial services. The design of the graphics, interiors and the printed matter of our industry was communicating messages about quality of product and service standards. Prudential's new design seems to have struck the right chord with the public, because we are now seen as contemporary and going places. Most different socio-economic and age groups find us appealing and this is particularly so with the younger middle class, exhibiting an efficient, businesslike approach which they regard as desirable. Some older groups thought the design had a certain ambiguity of gender, which intrigues me. On the other hand their overall impression of Prudential as being contemporary by our use of good graphics, attractive colours and typefaces is more important to them.

The other interesting difference came from the attitudes of the North and South. In the South reaction to the new design tends to be very favourable. In the North acceptance was somewhat muted. Some were unsettled by the speed at which we had changed. Could it be that we were being driven to change by our competitors? The older northerners were evidently concerned about sudden change, and I accept that it is human nature for some people to be cautious about coming to terms with a change. Is the South ready to accept change quicker than the North? I do not think that is true in all cases, but in our case it seems so.

And since I began by promising to draw you a picture, part of the research was to ask individuals to doodle on a piece of paper their impressions of companies 'in the recent past' and then again 'in the immediate future'. One doodle on Prudential's recent past turned out to be an elephant (shades of big and lumbering, I fear), but then the same doodler drew his impression of Prudential's future and this time he drew a space rocket on blast-off.

About eighteen months ago we defined the purpose of design in the Prudential as follows: to support and demonstrate visually our business style, which places value on simplicity, clarity, being contemporary, innovative and friendly. Because together, these values lead to quality in products and services, and design should reflect this quality.

This policy contributes to our business success in three different ways. As customers of financial products have become more discerning and questioning, their awareness of finance is growing. The new legislation gives them greater protection and places an obligation on the providers of financial services to be concerned with the customers' interests. It follows that the better the design to present our products, the more satisfied our customers are. The second way is the conditions under which our staff work. There are some people who claim that they go on working for the same reason that a hen lays eggs, but we have no wish to create a battery feeling at Prudential. Good design not only makes a working area more enjoyable, it also contributes to an increase in the effectiveness of work. And the third way is its effect on the general public, not all of whom are Prudential customers . . . although we do have rather a lot of them.

All our research results endorse that that design is working positively for us. To maintain a high standard we have a small team at the centre monitoring progress, providing advice, influencing and giving direction to the operational areas in their use of design. This is a design management group – not a group of designers – and they are the final arbiters in interpreting the corporate identity guidelines. It is their responsibility to see that prudence and what it stands for is not ravished by designers who would like to add their personal interpretation. I am not sure I should accuse designers of 'ravishing', but from our point of view the design is set for the moment, and any variation only diffuses the message.

I shall give you an example of one difficulty that arose: within a few months of our identity being launched we decided to move to laser printing technology. There were economic advantages in converting upwards of 20 per cent or more of our print from typesetting to laser printing. Detailed guidelines had been prepared for text setting using our typefaces. The design management team stipulated that the design processes for both methods must follow the same standards for layout and fonts. This would ensure visual consistency in the printed matter received by our customers, whichever form of printing was used. There were many technical objections to this. They have been overcome and so have the copyright issues. It would have been easy to agree with the laser printing specialists and accept the objections as insuperable. Potentially such an agreement would have visibly weakened – perhaps in time even destroyed – the identity.

Implementing the programme for design changes to product literature, uniforms, office interiors and signage takes time in a large a diverse company. Not only to set out the principles to achieve quality and consistency but to ensure that they are applied to a standard. Monitoring standards – policing if you like – is not easy in an organization managed under a devolved authority. Yet it is critical. Our identity and the inherent design qualities behind it are intended to draw together the diverse strands which make up the corporation and to present them under a common banner with a common purpose. This commonality of purpose as expressed by identity is not something I can delegate or relegate to business areas. The research I mentioned showed that this unity was appreciated by the customer. We need to ensure that our design management processes remain tight.

Let me finish with two points. First, the public is influenced by design. Secondly, it deduces from design impressions about the quality of goods and services. As a consequence, managing design, which in our case started out as managing a corporate identity, is a mainstream business activity which, with others, is critical to our future success.

*Sir Ralph Halpern is
Chairman and Chief
Executive of the Burton
Group plc. He is Chairman
of the Consumer Affairs and
Marketing Committee, a
Member of the President's
Committee of the CBI and on
the Board of the Governing
Council of Business in the
Community.*

Design for Profit

When I was asked to talk to the London Business School about design I reached for two things: a drink – and a dictionary. Under 'Design' I read: First. 'A plan or scheme conceived in the mind and intended for subsequent execution'. Execution sounded a trifle sinister, so I read on. Second. 'A project'. That was better. Brief and to the point. Third. 'A plan in art'. Now that I liked. It broadened the definition. Also, of course, design can be any number of things: a sign blowing in the wind outside a pub . . . how a room is lit . . . the way a woman's skirt hangs . . . how a restaurant presents its food . . . the shape, size, layout of a shop or store or brochure. Which brings me onto familiar territory, because for me design is, first and last, the designing of a business. A successfully designed business is one that has all its policies, strategies, tactics and management structure clearly defined and – a key point – in tune with the culture of the times.

Today's Burton Group with its 1700 outlets and over 35,000 employees began, like many a successful enterprise, with a man and his idea. The idea was simple enough. What mattered was that he was the first to think of it. Early in the 20th century Montague Burton gave men in their millions style and a modicum of elegance – scarce products after World War I – by designing a suit: a suit to fit their aspirations, their shape and their pocket.

Burton – which was almost his real name – was principally a manufacturer. I am largely a retailer – that is to say, a modern merchant. However, in our time we both observed a basic rule: before designing and producing anything, study the market and its needs. This is my theme: how a design-led attack on the market can produce success in business and, provided one is wide awake to a rapidly changing world, will keep you ahead of the game, and hopefully close to the top of your particular tree.

For years Montague Burton was enormously successful. His brilliant idea made his company. He led the field. However, time passed, fashion changed, investment in design faltered and his business increasingly failed to relate to a public that was growing more selective, more demanding and more prosperous. 'Only connect one with another', wrote the novelist E. M. Forster about the human race. Burton was no longer connecting. While he stood still the market that had sustained the business for years had moved. His products, shops, systems, factories, above all management strategies were designed for an era that was gone forever.

It is easy to be wise after the event, but if the essence of design is purpose and a clear direction, then the business that ignores those factors is inviting failure. Because Burton forgot that yesterday's and today's market, not to mention tomorrow's, are almost never the same, any product it made could only succeed by accident. Furthermore, a successive management had perceived design only in the narrower sense: something tacked on the end of the business, like a fresh lick or paint, and more often than not the product of a graphic designer's intuition.

What all this meant for my company was that the era of the specialist retailer had arrived. If our business was to respond, new ideas must be developed to target its most profitable sectors. Market analysis made three clear signals.

1.   The general retailer with a broadly-based store or shop which tried

to satisfy a cross section of customers of all types and all ages had had its day.

2. We should produce a number of chains of shops, each different, each organized in depth, which taken together added up to a substantial share of the overall market.

3. That market required precise and accurate segmentation, so that each business could be designed from the ground up, with a sharp focus on one particular segment and no other.

In addition, our customers were demanding added value through a form of clothing to fit a more casual way of life. In a nutshell, for Burton to act upon this change in the market was essential to its survival, and fundamental to that change in the market was design.

I had begun in the high street on a modest scale with Top Shop in the mid-60s. My team and I had concentrated on a single large section of the clothing market – young women between the ages of fifteen and twenty-five. Our business mission was to create a shop in which the only constant was change. To most people change is something they tend to avoid. What is familiar is what we feel safe with, even if it is outmoded and no longer functional. Old shoes are more comfortable than new ones, even if they leak, so we hang onto them and when it rains our feet get wet.

None of this applies to the young. They welcome change. They find experiment exciting. They love to startle. They find it stimulating and they find it fun. Our job was to manage change while remaining focused on our single-target market. The results were spectacular. We had tapped a vein that had not been tapped before. Like most people in this country I had instinctively resisted change. But now it dawned on me that the more change I was capable of accepting the greater the opportunities that would upon up before me.

Managing change is one of the most important skills that whoever runs a company can develop. And if we are to identify new horizons in a world that is constantly on the move, a primary task is to initiate and encourage change.

At Burton, then as now, design played a key role. Not tacked on, but the force which defined every aspect not only of Top Shop but of all that followed on from that first success. True, retailing is a highly visible activity where design is likely to be a priority. But I am talking about something more than that – what today is called 'total design'. From management structure and business strategy, down to clothes, stores, window displays, in-store layout and coat hangers, every aspect of our business was designed to meet the needs of our market. Even office furniture, stationery and the Christmas party reflected the business mission. The great advantage of this was that everyone knew exactly what they were doing and why. We knew our customers and what we had to do to satisfy them – and they knew us.

Whereas, in the past, management had failed to anchor its activity to a clear commercial purpose, our chain of Top Shops had one overall objective: to provide fashion-conscious girls with the clothes, accessories, atmosphere and prices they wanted and to do it all under one roof. It was their scene. No one else's. To appeal to them and their generation our shops were designed to use colour, lighting, music, flooring, anything that would say to the passing shopper 'come on in, the water is lovely'. The idea was to make the ordinary appear extraordinary. It was lively, it was original and it worked.

Of course, we made our quota of mistakes. One lesson we learned was that design does not exist in the abstract. It is not a smart word for appearance or an all-purpose gloss for the harder edge of a business operation. Design is a mood, a message, a frame that sets off the picture: if you like, a form of theatre. Whatever business one happens to be in – retailing, transport, engineering, technology, the leisure industry – design is integral to the generation of profit. It is what differentiates you from your competition.

Many famous British companies have demonstrated this with outstanding results: IBM, Jaguar Cars, Vickers, British Airways and many, many more have secured a place in the affections of huge numbers of people by designing their corporate plan, management strategies, company tactics and customer relations around precisely what it is their customers want and expect, and then providing it.

From Top Shop the good word spread. We had found our market and we became the best-known fashion chain in Europe. Indeed we were at that time the only section of The Burton Group that was making a profit. Which explains why, one day in 1977, I was invited to become the company's chairman and chief executive. I accepted the challenge, albeit with some misgiving, and took my Top Shop team with me. Most of them are still with me to this day.

We were ambitious. We knew where the future lay. Segment the total clothing market, analyse each segment in depth and then design new businesses to satisfy them. Having done our homework we set out to broaden our horizons, to become Britain's most successful fashion retailer. We knew it would take years but never mind, that was our target.

We began by launching a design-based attack on the largest and most profitable sectors of the clothing market. And so out of Burton and Top Shop were born Top Man, Evans, Principles for Men and Women, Champion Sport and acquisitions like Dorothy Perkins and more recently Debenhams, of which more later.

In all this, design in both the narrower and broader sense played a central role, constantly signalling the business mission to whatever age group we had in our sights. In the past designers were traditionally decorators and as such would often add decoration unrelated to the product. Then, about twenty years ago, there was a change. A businessman would say: 'I must build a shop of design a product. I shall get my designer to develop a personality for me with a nice fascia or carrier bag that will make the customer recognize me better than anyone else and that will help to make me profitable.'

However, then someone came along and said, 'Just a moment. There is one thing called merchandising and it is important. First, get your customers inside the store; then, once in, persuade them to move from one part of the store to another so that hopefully they will purchase more than they originally intended. Do this by using various merchandising methods, but do not do it in a conventional way, make the whole environment interesting and amusing and enjoyable.'

People used not to enjoy shopping. Especially the young. They either hated it or were bored by it. They did it because they had to get, say, a pair of shoes, because you could not go around barefoot, and it was not much fun. It was literally a bare necessity. Today, with the help of our designers, we have made shopping a lot more fun.

At this point I should like to make clear that designers do not just pop out of the woodwork, waving plans for a great new store. They must be briefed. It avoids misunderstanding if the chief executive – that is, the chief designer of the business – does the briefing himself, explaining what markets he is going for; the kind of sales per square foot he hopes to make; the amount of stock he is going to carry; the staff, computer and customer services he needs and so on.

There is one thing you do not say to a designer. 'Do you see that store over there? I am thinking of buying it next week. Go and give it the right look.' If you do, you will get the wrong look designed for the wrong market, or, more likely, no market at all. A shop with flashing lights, a young ambience and pop music going flat out is not likely to attract your middle-aged customer, except perhaps as a curiosity. Furthermore as people get older they find a way of living that suits them and they tend to stay with it. And a more calming influence is what they look for.

Regardless of the most detailed brief, there are of course designers who are not commercial. They design solutions that do not sell because nobody quite knows what or whom they are for. For example, do you remember the Sinclair C5? What was it? A sophisticated bicycle? Fun on wheels? Or a new sort of Mini? The fact is, design is indispensable and its effect immediate. But without clarity of purpose the most skilfully constructed product can fall flat on its face. When that happens, design is devalued, old mistakes are remembered, and people shake their heads and mutter.

To take a couple of examples, tower blocks are not exactly God's gift to the human race, either aesthetically or to those who live in them. Then there is the mud-coloured exterior of our National Theatre on the South Bank, likened by the Prince of Wales to the blueprint for a nuclear power station. At the same time, on the other side of the ledger there is the new Mound Stand at Lord's with its splendid canopy that sets off that famous ground to perfection. Then there is the simply brilliant – and brilliantly simple – logo of the white mask that covers the Phantom of the Opera's face, unmistakably inviting the world to 'come in and be scared to death'; an invitation that the world continues enthusiastically to accept.

We have first-class designers in this country – bold, imaginative and highly skilled. There is more than enough work for them. But it is up to us businessmen to make sure we use them properly. To this end Burton has sponsored courses at the Royal College of Art, to help bridge the gap between businessmen and designers; to encourage them to stay here and become part of our business community rather than vanish overseas – a story that used to be all too familiar.

Design has also been given a boost by the present government, which has banged the drum for individuality and self-expression, and since 1979 created a climate in which it can flourish. It is interesting to note that Burton's success has coincided with the revival of our country's fortunes. Before then people's confidence tended to be subdued. They were not encouraged by the government of the day to express them-selves individually but rather through some communal activity, working with others: such as being a part of British Rail, perhaps, or in the civil service, or as a member of the banking community. Thoroughly repu-table and necessary jobs that stood for something, but in which the employee was more a number than an individual. Of course, there were exceptions but by and large that was the picture. The state looking after you was the order of the day. Individuality in business and behaviour was at a discount.

Likewise individuality in dress. People tended to mass together and 'follow my leader'. The 'mini' was 'in' so every girl wore a mini. If 'turnups' were 'in', everyone wore turnups. If it was lapels . . . and so on. We lurched from one fashion trend to another. Oddly enough, that was the beginning of fashion taking a high profile. However, people wore the same thing. It was all for one and one for all. Conformity was 'in', individuality 'out'.

Today, what a change is here! This may or may not be the dawning of the age of Aquarius, it certainly is the age of the individual. And, as ever, fashion reflects the political dimension. Look around you and you see through clothes and hair, individuality flaunting itself where yesterday it would have caused ribald mockery if not instant arrest. Today, if anyone raises an eyebrow, it is in admiration.

Thus design is high on the hog, as people develop their individuality in a variety of ways. In today's Britain there is variety in everything we touch: variety in fashion, variety in houses, variety in cars, variety in food. Especially in food. As we move towards the classless society, all sections queue at the fish and chip shop. Then, while there is still roast beef and Yorkshire pudding, there is also the hamburger, the pizza, the Indian or Chinese take-away, the Greek or Turkish kebab, the Mexican tacos. Wherever you turn you find the British opening out, eager to sample the international market that is right here on our doorstep. It is a social trend that is ready-made for the designer.

If by any chance there were a reversion to a period of austerity, individuality would, I think, once again be at a discount. There would be little demand for creative design because people would be unobtrusive again, head once more below the parapet, in a socialist state where equality is the name of the game. A good example of this is China, the recent students' revolt notwithstanding. In China everyone dresses the same from the chairman of the party down to the worker on the railway. And even under Mr Gorbachev the mass of the Soviet people are still in the very early stages of self-expression.

Here in Britain the emergence of the designer reflects the opposite: a society which encourages individualism with the people shopping around, picking and choosing and welcoming choice. Will choice go on? As long as individuality, wealth creation and the independent spirit go on. Nevertheless in the face of this individualism the most recent trend is that all the large retailers are moving towards a centralist theme. Sameness is – in every sense – all over the shop: in design, in production, in the High Street environment. The wheel has come full circle. How Monty Burton must be laughing.

It has to do with a computer-led society, where the computer all but tells you what to buy and what to sell. And because of that we are all buying and selling similar things. White sells best, so let us have thousands of white shirts. You would rather have blue, madam? Try next door. Or over the road. They have only got white, too? Now there is a coincidence. I think there is going to be a tremendous need for retailers to make up their mind to be different again, in a fresh, ingenious way that does not mean people going back to being bored with shopping. As ever, design will be the most important part of the process. Someone will lead it, but there is not a clear leader at the moment because every large company in my field is heading in the same direction, all looking alike and selling alike.

Let me give you an example. When we acquired Debenhams about five years ago we redesigned the whole business – stores, strategies, markets, merchandise and presentation, following what the market told us. We implemented the Galleria Design Concept for the Oxford Street store; an idea which caught everyone's imagination and has entered the standard business dictionary. The competition watched, hoping – well, it is natural enough – hoping we would fail. However, now it is clear that we have brought a whole new shopping experience to the high street. Already six new stores, unique in concept and design, herald the first and only speciality chain in the country that is dedicated exclusively to individual and home fashion. The choice is wide, with a range that is selective and not to be found in the average department store. So now they are all doing what we did – editing out the ranges, the merchandise that was not appropriate, and focusing on fashion for the home, having pianists on the ground floor and reshaping their stores in our image. One of these days you will not know if you are in Debenhams or in Dickens & Jones.

When everyone rushes onto a bandwagon like that, the Pulitzer Prize goes to whoever comes up with the best selection, best service, best quality, best style at the best price. Thus the next phase is product differentiation. That is the phase the high street is heading for – and in that design will be right up front.

So much for the immediate future. What of the long term? Given imagination the opportunities that lie ahead, such as Sunday trading (which will happen, as will shopping by television), the opportunities are almost limitless. So are the questions they pose. But in some ways the keys to success will remain the same.

From the early Top Shop days I have tried to design the whole business, whether at individual, chain or group level, around a clearly understood purpose, a design with direction, with content and with form. In a competitive world we must be constantly thinking up new things to do and new ways in which to do them; leap-frogging over the present and reaching out to the future which belongs, as it always has, to those who see possibilities before they become obvious. I hope no one will think it presumptuous if I suggest that what we began with Top Shop paved the way for The Sock Shop, Body Shop and Tie Rack.

One of the problems of the affluent 80s – and we are an affluent nation whether we acknowledge it or not – has been our habit of focusing on more of what has been wrong and unfinished than on what is right and progressing. 'There exists in human nature,' wrote Edward Gibbon 200 years ago, 'a strong propensity to depreciate the advantages and magnify the evils of the present times'. Indeed it is a certainty that if we perfect an ideal design for living we shall be everlastingly happy. The problem is that every solution creates new problems. Good. That gives people who design for tomorrow a chance to create new solutions.

For example, we look about us and we see the growing brutalization of our lives, and in particular those of our children. The manufacturer and the merchant are well-placed to influence youth. It would perhaps be no bad thing for some of us to take the lead in designing for the young a more harmonious and attractive environment, as a counterweight to the crude, the ugly, and at times the mindless.

So let me end with the unforgettable creed that Shaw gave the dying artist in *The Doctor's Dilemma*:

'I believe in Michelangelo, Velázquez, and Rembrandt; in the might of design, the mystery of colour, the redemption of all things by beauty everlasting, and the message of art that made these hands blessed.'

Ladies and gentlemen, we may not reach Valhalla, but let us at least never cease to strive towards it.